WASTE AND WASTE MANAGEMENT

FOOD WASTE

PRACTICES, MANAGEMENT AND CHALLENGES

WASTE AND WASTE MANAGEMENT

Additional books in this series can be found on Nova's website under the Series tab.

Additional e-books in this series can be found on Nova's website under the eBooks tab.

WASTE AND WASTE MANAGEMENT

FOOD WASTE

PRACTICES, MANAGEMENT AND CHALLENGES

GARRETT LEONARD RILEY
EDITOR

nova publishers
New York

Additional color graphics may be available in the e-book version of this book.

Library of Congress Cataloging-in-Publication Data

Names: Riley, Garrett Leonard, editor.
Title: Food waste : practices, management and challenges / [edited by] Garrett Leonard Riley.
Description: Hauppauge, New York : Nova Science Publishers, Inc., [2016] | Series: Waste and waste management | Includes bibliographical references and index. | Description based on print version record and CIP data provided by publisher; resource not viewed.
Identifiers: LCCN 2016018870 (print) | LCCN 2016011332 (ebook) | ISBN 9781634850483 () | ISBN 9781634850254 (hardcover) | ISBN 9781634850483 (Ebook)
Subjects: LCSH: Food waste. | Organic wastes. | Refuse and refuse disposal.
Classification: LCC TD804 (print) | LCC TD804 .F67 2016 (ebook) | DDC 628.4/45--dc23
LC record available at https://lccn.loc.gov/2016018870

Published by Nova Science Publishers, Inc. † New York

CONTENTS

Preface		**vii**
Chapter 1	Food Waste: An Expression of the Evolution of Current Agricultural Development Systems *Simone Vieri and Grazia Calabrò*	**1**
Chapter 2	Food Wastes as Sources of Natural Antioxidants *Adriana S. Franca and Leandro S. Oliveira*	**23**
Chapter 3	Potential of Spent Coffee Grounds as Sources of Dietary Fiber with Antioxidant Activity *Leandro S. Oliveira and Adriana S. Franca*	**51**
Chapter 4	Potential Applications of Food Wastes to Bioenergy by Anaerobic Processes *Sandra I. Maintinguer and Lorena Oliveira Pires*	**71**
Chapter 5	Anaerobic Digestion: A Promising Solution for Food waste Management in Urban Areas in Vietnam *Hoa Huu Nguyen*	**95**
Chapter 6	Measuring Food Waste Generation from Households in Greece: Behaviours, Attitudes, and Potential for Prevention *K. Abeliotis, C. Chroni and K. Lasaridi*	**121**
Chapter 7	Safety Analysis of Soy Sauce Squeezing Residue and Fish Meal Storage by Thermal Analysis and Gas Chromatography *Naoharu Murasawa*	**137**

Chapter 8 Sugarcane Bagasse Hemicellulose Properties, Extraction Technologies and Xylooligosaccharides Production **155**
Michel Brienzo, Ana Flavia Azevedo Carvalho, Franciane Cristina de Figueiredo and Pedro de Oliva Neto

Index **189**

PREFACE

An increased demand for recycling has prompted the food industry to become more efficient in its handling of waste. Food waste is among the priority streams for waste prevention worldwide, as it is generated at large quantities at all levels of the food production and consumption chain, in both developed and developing countries, throughout the year. This book discusses practices, management and challenges of food waste. The first chapter discusses current agricultural development systems. Chapter Two presents a review of the works of research that have been developed in the last decade on the use of food wastes as sources of antioxidant compounds, such as polyphenols. Chapter Three provides a critical overview on the studies that deal specifically with food applications, mainly focusing on their potential as sources of fibers and associated phenolics and respective antioxidant activity. Chapter Four contributes to innovation in research on the valorization techniques that have been developed for the use of food wastes as well as examines future directions in the field. Chapter Five reviews conventional methods currently applied in Vietnam for treatment of food waste, and then proposes a promising solution using anaerobic digestion technology. Chapter Six provides an overview of the generation of food waste within households in Greece and investigates the potential for prevention. Chapter Seven investigates heat generation and oxygen levels during the storage of soy sauce squeezing residue and fish meal, by-products of soy sauce and fish production and processing, respectively, by thermal analysis and gas chromatography. The final chapter discusses sugarcane bagasse hemicellulose properties, extraction technologies and xylooligosaccharides production.

In: Food Waste
Editor: Garrett Leonard Riley
ISBN: 978-1-63485-025-4

Chapter 1

FOOD WASTE: AN EXPRESSION OF THE EVOLUTION OF CURRENT AGRICULTURAL DEVELOPMENT SYSTEMS

Simone Vieri[1,*] and Grazia Calabrò[2,†]
[1]Department of Management, La Sapienza University of Rome, Roma, Italy
[2]Department of Economics, University of Messina, Messina, Italy

ABSTRACT

Though a global problem, food waste has different intensity and modalities depending on how developed the agrifood systems of the interested countries are: in developing countries, the phenomenon concerns mostly the loss of agricultural products, while industrialized countries are mainly interested in food waste. Loss and waste make up the main concerns of today's development model, and are to be considered the expression of inefficiencies and distortions of processes of production and distribution, the costs of which tend to weigh on the weaker members of agrifood supply chains - farmers and consumers. The objective of this paper is to consider the issue of waste not as a single unrelated problem, but in the wider context of the recent evolution of the current models of agrifood development. In this way, we can highlight the main basis that lead to the inefficiencies that underlie loss and waste,

* Corresponding author. Email: simone.vieri@uniroma1.it.
† Email: calabrogr@unime.it.

and subsequent economic and social costs. More specifically, waste will be considered in relation both to the distortions of the competition regime, set by the great differences in bargaining power between the various members of the agrifood supply chains, and in relation to the main issues of the current agrifood development models, mainly the issues of malnutrition and environmental sustainability.

Keywords: food waste, food losses, food products, agrifood supply chain, agrifood systems

INTRODUCTION

Losses and waste of farming products and food which occur, for various reasons and in various ways, in the agrifood supply chains, are to be considered an obvious expression of inefficiencies which typify all the various production steps of the supply chains themselves. In fact, it is obvious that the costs brought about by losses and waste should not only be referred to the value of the interested products, but also to the environmental and social problems they cause.

Although the issue of waste has been known a long time, the attempts to quantify it are very recent. In this sense, the first and most significant attempt has been that of FAO in 2011, with the publication of a global report (FAO 2011), followed by further studies to quantify the economic and environmental impact of food losses and waste.

Already in the early 80s, the FAO had detailed a definition of food waste:

<<Wholesome edible material intended for human consumption, arising in any point in the food supply chain that is instead lost, degraded or consumed by pests>> (FAO, 1981).

Later on, such definition was honed (FAO, 2011), including a distinction between food losses and food waste. According to this definition, food losses are those that take place at production, post-harvest and processing steps; food waste, on the other hand, occurs during distribution, sale and final consumption (Lipinski et al., 2013).

According to the FAO, food loss must not only take into account quantity, but also loss of nutrients. In this sense, a more advanced concept compared to the previous one is that: <<Food loss is defined as the decrease in quantity or quality of food and are the agricultural or fisheries products intended for human consumption that are ultimately not eaten by people or that have

incurred a reduction in quality reflected in their nutritional value, economic value or food safety>> (FAO, 2012a).

Another definition of the complex phenomenon of food losses and waste is that developed by the United States Department of Agriculture (USDA) which considers food waste a component of the wider problem of food loss. Particularly, food loss is considered <<the amount of edible food, postharvest, that is available for human consumption but is not consumed for any reason; it includes cooking loss and natural shrinkage (e.g., moisture loss); loss from mold, pests, or inadequate climate control; and plate waste>>; whilst food waste <<occurs when an edible item goes unconsumed, such as food discarded by retailers due to undesirable colour or blemishes and plate waste discarded by consumers.>> (Buzby et al., 2014).

According to 2011 FAO data, waste and losses amount in total to 1,295 million tonnes, equal to about 30% of global production, and are thus divided: 32,0% in the farming step; 46,3% in post-harvest; 21,7% in the consumption step.

The total of losses and waste in yearly pro-capita quantity are higher in more advanced countries than in less advanced ones: about 280-300 kg/year in Europe and North America; 120-170 kg/year in Sub-Saharan Africa and South/South-East Asia. This data, though expected, gains significance if compared to the pro-capita quantities of food destined to human consumption in the same geographical areas: about 900 kg/year in North America and Europe, 460 kg/year in Sub-Saharian Africa and South/South-East Asia. This means that, on average, losses weigh very differently on food produced, both in rich countries (from 31% to 33%) and in poorer ones (from 26% to 36%).

The stark difference is the pro-capita quantity of food waste generated by end consumers, estimated to be between 95-115 kg/year in well-off countries and between 6-11 kg/year in poorer ones. Confirmation of this is the fact that, in developing countries, over 40% of losses are concentrated in the post-harvest and early processing step, whilst in rich countries the same percentage applies to waste occurring during retail and final consumption. As a consequence, the FAO estimates that consumer food waste in industrialized countries is slightly lower than the food produced by Sub-Saharan African countries (222 million tonnes vs. 230 million tonnes).

The fundamental difference of the problem of food waste in rich countries and developing ones greatly influences the value of the interested products, evidently higher for food waste than for loss of primary products.

According to the FAO, the losses and waste that occur in the various steps of the food supply chains have an annual cost of about 750 billion dollars, calculated on the basis of the prices at production of the interested products. This method of calculation, as highlighted by the FAO itself, leads to a negative estimate of the real values, since they tend to increase along the supply chain, and therefore cannot be calculated using the price at production.

To try and make a better economic quantification of waste, specific studies have been made on the various conditions in each single country. In the UK, for example, it was estimated that the value of products wasted varies from 1,500 dollars to nearly 3,000 dollars per tonne, depending on whether the waste occurs during processing or during final consumption (WRAP, 2013). A similar conclusion was drawn for the USA (USDA, 2014) where waste was estimated to be of 2,700 dollars per tonne.

However, for retail prices – not prices at production as considered by the FAO – of the products interested by losses and waste, the estimated global cost (WRAP, 2015) is higher than that estimated by the FAO (750 billion dollars) by at least 400 billion dollars a year, calculated on the estimated value per tonne of food wasted: $2,500 for USA, Canada and Europe; $1,000 for industrialized Asian countries: $550 dollars for all other geographical areas.

Finally, as far as the single categories of products are concerned, according to the FAO (FAO, 2012), most losses and waste occur for roots and tubers and fruit and vegetables, reaching 45% of total production, followed by fish and seafood with 35%, and cereals with 30%; for dairy, meat, legumes and oil seeds, losses and waste are around 20%.

Though the problem of losses and waste can affect other important issues such as food access and environmental sustainability due to its sheer volume, it is not to be considered a cause, but rather the effect of the current agrifood systems' criticality. Viewed in this light, the problem of losses and waste becomes part of a particularly complex picture. Its typical elements are the evolution of the relationship between the various components of the agrifood supply chains, and the consequent obvious inefficiencies and distortions which determine the environmental, social and production problems of which losses and waste are a clear expression.

Waste and Dominant Positions in the Agrifood Supply Chains

Since after the Second World War, agriculture developed focusing on increasing productivity, which in turn was an expression of the intensification of the relationship between socio-economic components. An example can be the growing dependence of farming's upstream and downstream sectors on, respectively, purchase of production factors and transfer of products; another is the reduction of labour and of use of land for farming, which has accompanied socio-economic development processes everywhere.

The increase of agricultural productivity has undoubtedly served to meet the food requirements of an ever growing world population, ever more concentrated in cities and ever less employed in agriculture. However, there is no doubt that the agro-industrial models that have been employed up to today to support agricultural development have not only determined a strong increase in productivity, but also deep alterations of agriculture's typical characteristics (Anderson, 1987; Ruttan, 1984). In particular, agriculture has been forced to comply to industrial logic, simplifying its activities and production techniques, which more and more often push towards a single product and a single way to farm it. It is no accident that one of the main characteristics of agro-industrial models is that they are founded, all over the world, on monoculture, which in turn is characterized by great uniformity, both for factors used and for products obtained. Because of this, agriculture has been forced to become decontextualised from its environment and to give up its main characteristic and need, i.e., to base its production result on a balanced use of available resources, not exploitation (Vieri, 2012a). Thus it has been unavoidable that where farming has complied more strictly to industrial conduct, it has also ended up as being more subordinated to the objective of profit of the sectors it submitted to, and be less and less able to point its development on the basis of its needs and characteristics (Vieri and Calabrò, 2014a).

All the above has occurred because the physiological economic decline of agriculture in advanced economic systems has been determined by various factors: firsts among these is the tendency to concentration, particularly strong in industry and distribution, and practically null in agriculture.

There are various realities that coexist within agrifood supply chains, which only in agricultural production can be likened to pure competition. The situation is very different for the market of production factors and end products, where concentration has often be such as to lead to the creation of oligopolies (Vieri, 2012b).

To prove it, it is enough to consider that today, at a global level, 90% of the seed market and 80% of the agrochemicals one is controlled by six multinationals (Phillips McDougall, 2014); more than 30% of the market of food products is in the hands of 15 large-scale retail channels; about 20% of packaged food products is produced by 20 food industries (WRAP, 2015).

In this frame, a drastic weakening of the spread of development is inevitable, since the progress made (in terms of productivity, for example) by the various oligopolistic groups tends to be kept within, in the form of higher profits, instead of being spread outside in the shape of price reduction of production factors or end products. Therefore it is normal for there to be distortions which, in turn, generate inefficiencies and waste. As a consequence, inequalities grow, since the costs of such distortions are always borne by the weaker subjects, in our case on farmers and consumers. There is no doubt that towards the upstream and downstream sectors farmers are price takers. It is also clear that the cost of losses and waste is included in the price of products purchased by consumers.

A situation of this kind, which sees subjects with very different bargaining power within the same agrifood supply chain, makes it very difficult to put together and apply an efficient action against waste. Such an action, to be effective, should involve all parties within the various steps of the supply chains themselves. This is mainly for two reasons:

- The various subjects making up the supply chains have differing strategies, objectives and economic weight, and behave not as parts of a single system, but as "parties," whose main objective is to look after their own interests, each in virtue of their bargaining power;
- The creation of inefficiencies and the placing of its costs on the weaker members of the system is typical of oligopolies, which are the prevailing market form in capitalist economies.

Due to the above, the situations that seem to better adapt to the chance of applying waste cutting programmes are those with agrifood systems dominated by large groups that deem convenient to reach that objective. In Europe, for example, the main initiatives to cut waste have been started either by companies that hold a large stake of the food markets in the interested countries, or by governments of countries with agrifood systems with a prevailing number of large groups or well-organized representative offices. Some examples are the initiatives of a multinational in France operating in large-scale foodservice (BCFN, 2012), of some groups operating in large-scale distribution and public institutions in Great Britain (WRAP, 2011; WRAP 2013), or of a group of food industries in Norway (Hanssen and Møller, 2013);

programmes have also been launched by governments in the Netherlands (Rutten et al., 2013) and Denmark (SWF, 2013).

It is more difficult to launch anti-waste initiatives in countries with fragmented production systems, with many small and medium-sized companies operating in the various steps of the agrifood supply chains, making it difficult for them to be coordinated. As a consequence, it is even more complex to act in developing countries where the main problem is not waste, but losses, and where, to curb such phenomenon, it would be necessary to act on infrastructures and production structures.

It is ironic that, at least in more developed countries, one of the main causes of distortions and inefficiencies causing waste, i.e., the development of oligopolies, becomes one of the best tools to curb the problems it has created. This should spur reflections on the limits shown by the current agrifood models and, more generally, by the current models of development. In particular, it would be fit to reflect on the difficulties to even hypothesise alternative models to the current ones, and how they can be considered congruent with every peoples' right to food sovereignty.

Waste and Other Paradoxes of the Agrifood Systems

Among the main criticalities of agribusiness production systems one is certainly the inefficiencies in the use of resources and distribution of products. In this scenario, the issue of waste and losses is important in determining the main problems of the system, such as sustainability, malnutrition and the efficiency of production processes. It follows that the problem of waste contributes to, and at the same time is a direct expression of, what is, in all likelihood, the gravest paradox of our times, where 38% of the Earth's population suffers of food-related problems (too much or too little), about 40% of agricultural production either becomes animal feed or has a no-food use, about 30% of food is lost or wasted (Vieri and Calabrò, 2014b).

Taking the above into account, one of the first aspects to consider is that, to be produced, food products lost or wasted need land, water, capital, labour and, more generally, power, just like products that are not food (FAO, 2013a).

According to the FAO, about 28% of farming land is used for crops whose products go lost. It is a huge amount of farming land; if it were a single country, it would be the second largest country in the world. This in itself is critical enough, but is made more so by the fact that in the world every year, about 12 million hectares of land are lost due to various phenomena of depletion (FAO, 2015a), an area not much smaller than the farming land used in a country such as Italy.

Over the last 50 years, in the world, agriculture has trebled the quantity of production, with an increase in farming land of just 12%. This means the exploitation of soil has been greatly increased, so much so that intensive farming is among the main causes of its depletion. It must be added that, for the long time necessary to its formation and considering our life-spans, soil is to be considered a non-renewable resource. Loss of soil is estimated to be at levels higher than 30 times the sustainability rate (Pimentel et al., 1993).

It follows that the use land for crops whose products go lost is to be considered not only in view of the use of soil, but also in connection to the kind of agriculture practiced and its soil depletion effect, since it is highly likely that the consequences of the two phenomena (loss of products and of soil) add up. It is plain to see that to use land for crops whose products go lost, especially in intensive farming or farming that is not sustainability-oriented, brings on the added damage of contributing to loss of land that could have been avoided.

Similar considerations can be made concerning the use of water. According to the FAO's data from its water information system (Aquastat, 2016), agriculture is the sector in the world that draws the greatest quantity of fresh water: 69%; 19% is used by industries and 12% by municipalities. Such levels of consumption are particularly high (over 80%) in Africa and Asia, where there is also the greatest waste due to the predominance of surface irrigation. Of 324 million hectares irrigated today, 280 million use surface irrigation; 29% of irrigated land in the world is rice fields; 42% of irrigated surfaces are in China and India. Furthermore, to meet the yearly water requirements for world agriculture, measured in about 1,500 km^3, 2,700 km^3 are used, with an average efficiency of 56%, with positive peaks of 72% in North America and 70% in Russia, and negative peaks of 28% in Sub-Saharan Africa and 26% in Central America.

In this worrying scenario, the issue of resources and waste bears heavily, since according to the FAO, it is estimated that about 9% (250 km^3) of water used in agriculture is destined to crops whose products end up lost or wasted. In this case, as for soil, if we were to ascribe to a single country the use of water for crop irrigation, whose products go wasted, that country would be the first in the world for use of water for irrigation. It must be highlighted that in the world, of all the surfaces irrigated, 61% are cereal crops, 10% vegetables and 6% fruit, and that these sectors register losses of 45% (fruit and vegetables) and of 35% of productions obtained at a global level. In this case too it is clear how the loss of farmed products means waste of water that could otherwise have been saved or used differently.

Considerations on the capitals used in agricultural and food production processes must be added to those on the waste of natural resources. An interesting element for synthesis can be the power used to obtain agricultural products and food (FAO, 2012b), which is equal to about 35 gigajoules per person per year in OECD countries, and to about 8 gigajoules in developing countries. It is to be noted that half of the abovementioned power consumption is used in the step of industrial transformation and distribution in OECD countries, and for cooking in developing countries. As a whole, agrifood supply chains absorb about 30% of world power, of which 70% is used in the steps following the farming one. It is estimated that, together, losses and waste absorb about 38% of the total power used by agrifood supply chains (equal to 11.4% of power available at a global level).

With reference to chemical substances used in the production process, and therefore to production factors which, in the case of lost or wasted productions, represent a useless release of pollutants in the environment, it is to be noted that both the use of pesticides and of fertilizers has seen an increase. Particularly, from 1990 to 2014, the pesticides market has gone from little less than 46 billion dollars to over 55 billion dollars, and a further increase is estimated until 2018 (+2,6% a year), when it should reach 62 billion dollars. The main market shares (about 25%) and consumption are concentrated for over 50% in Asia and Latin America (Phillips McDougall, 2014), areas with a significant loss of farmed products.

The use of fertilizers has significantly increased (FAO, 2015b). The use of the overall three macronutrients (nitrogen, phosphorous and potassium) has gone from 161.8 to 190.7 million tonnes from 1990 to 2014 (+19.1%), and a further increase is estimated (+5.1%) until 2018, when it should reach 200.5 million tonnes. It must be stressed that the estimates on the increased use up to 2018 are particularly high in those areas with the greatest losses of farmed products, meaning Africa (+3.6% a year) and Asia (+3.1% a year), with peaks in Sub-Saharan regions (+4.7%) and South-East Asia (+4.9%).

Besides the abovementioned aspects, the effect of losses and waste on the sustainability of agrifood supply chains' production processes must also be considered concerning greenhouse gas (GHG) emissions. The FAO estimates that every year, due to losses and waste, about 3,3 million tonnes of CO_2 equivalent are produced, meaning a quantity of GHG lower only to that released by China and the USA (FAO, 2013b). If we were again to consider the total emission of greenhouse gases deriving from losses and waste of food products as made by a single country, this country would be the third in the world. It is to be noted that the quantity of emissions varies greatly from product to product, according to what step of the supply chain it is produced by, and to the production and food systems of each country. As an example, the quantity of emissions for vegetable products tends to double in the passage from farming to retail sale; for the same steps for red meat the increase of emissions is estimated to be just 4%; however, the emissions ascribable to the production of a tonne of vegetable products is much lower than that ascribable to the production of the same quantity of red meat (WRAP, 2015). Similar considerations can be made for the single countries in relation to their main productions and consumption.

Another significant aspect of the current paradox of agrifood systems is that pertaining to malnutrition, meaning the overall criticality concerning human nutrition, which goes from scarcity of food and micronutrients to excess assumption of food, and therefore problems linked to overweight and obesity.

As far as undernourishment is concerned, the problem of losses and waste is relevant for all the four basic conditions for food security, i.e.: availability, access, stability, utilization.

As Nobel Prize winner Amartja Sen wrote: "Starvation is the characteristic of some people not having enough food to eat. It is not the characteristic of there being not enough food to eat" (Sen, 1981). This is confirmed by a level of availability of food which, statistically, is able to satisfy global needs. FAO data on food security (FAO, 2016) shows that in 2014-2015, in every area of the world, the index of the "Average Dietary Supply Adequacy" is over 100%: 122% globally; 136% in developed countries; 120% in developing countries; 111% also in Sub-Saharan Africa, despite the fact that in this area are concentrated 27.7% of the undernourished people, and where their number has risen from 175.7 million in 1990 to 220.0 million today.

In this context, to better understand how losses affect food safety in poorer countries, it is useful to link them to the most important products in the diet of the interested populations. For example, the FAO "Share of Dietary Energy Supply derived from Cereals, Roots and Tubers" index highlights that in Sub-Saharan Africa 64% of the energy supply of the population depends on such products. In this case it is inevitable that losses which register their highest peaks for these products and in this geographical area greatly affect the food security of these populations.

As far as malnutrition due to excess food is concerned, it must be highlighted that they are concentrated in developed countries, where the most relevant problem is waste, especially in the final steps of the agrifood supply chains, i.e., retail sale and consumption. Given the growing significance of the problem, excess consumption of food and its effects, in terms of overweight and obesity, cannot be considered the result of individual problems with food, rather as a true system distortion. According to FAO data, the "Prevalence of Food Over-Acquisition" index shows that the percentage of individuals who tend to intake excess food regularly compared to their need is growing everywhere: at a global level, from 1990-92 to today (2014-16) the number has grown from 23.0% to 30.8%; in developing countries from 178.7% to 27.6%; in developed countries from 38.8% to 45.7%. The situation is particularly serious in countries like Belgium, the USA and Canada, where the percentage of people with a regular intake of more food than necessary is 66.2%, 60.8% and 58.7% respectively. It is clear that we are faced with a complex problem that shows how widespread the culture of waste is, since it is induced by consumerist culture which, by its very nature, can only contribute to more waste, even in agrifood supply chains.

Initiatives Against Losses and Waste

Because the problem of losses and waste occurs with greatly differing intensity and modalities due to various factors, which in turn are linked to the socio-economic characteristic of the single interested countries, it is very difficult to organize and apply a single strategy. Rather, strategies should be fashioned according to the various realities they must tackle.

For this reason in May 2011 the FAO launched "Save Food: Global Initiative on Food Loss and Waste Reduction," highlighting the need to set up and apply specific programmes in cooperation both with other international organizations and national governments, involving all the parties of the agrifood supply chains, from farmers to multinationals.

It is in this spirit that the FAO launched the initiative Save Food, in cooperation with Messe Düsseldorf, the largest organizers of trade fairs in the world. Already in its first stages it has been joined by 250 partners – organizations and public and private companies. However, because of the number of parties involved, its approach - obviously different on the basis of the socio-economic characteristics of the interested geographical areas - needs to be multidisciplinary and integrated. Because of this, Safe Food will use and apply different strategies for different regions, having pinpointed eight "Save Food Regions": the European Union, North America and Australia, Japan and the Republic of Korea, Eastern Europe and Central Asia, North Africa and the Near East, Sub-Saharan Africa, South and East Asia and the Pacific, Latin America and the Caribbean.

As far as the FAO is concerned, actions to fight losses and waste are also part of more general strategies, such as those for sustainable development and the fight against malnutrition. It must be said that the objective to eliminate food losses and waste is one of the five goals of the "Zero Hunger Challenge" programme launched in 2012, and is among the 17 Sustainable Development Goals identified by the document "Transforming Our World: the 2030 Agenda for Sustainable Development" adopted on 25 September 2015 by the United Nations' General Assembly for a new development policy favouring poorer countries.

A global vision and the cooperation between private and public parties to apply integrated actions differentiated on a regional base are undoubtedly a significant premise, but are not decisive to overcome the complex phenomenon of losses and waste. In fact, such problem must be considered mainly an effect of the distortions which characterize agrifood supply chains and of the difficulty to make different actors with different characteristics and economic weights converge on common objectives.

This also makes the governments' work very arduous. Without active cooperation from private parties, governments can't reach results that go beyond futile awareness campaigns. Such a situation contributes to making the fight against food losses and waste even more complex and geographically difficult.

In fact, it is obvious that active participation of private parties to campaigns to reduce waste is inevitably linked to the possibility of creating economic interests on which they can converge. For developed countries, it is also clear that such a convergence is easier with the cooperation between governments and large companies that operate in the various steps of the supply chains (production factors suppliers, processing industries, large-scale distribution), i.e., parties who, because of their dominant role on the market, are also the most accountable for inefficiencies which later bring on waste, among other things.

The situation in developing countries is even more complex, since the main problem is food loss. Here actions need investments in structures and infrastructures, and it is therefore particularly difficult to find the right convergence of public and private interests. In many cases, the investments necessary to curb food losses can be likened to investments in cooperation in development, a sector that, particularly these last years, has been at the centre of heated debate concerning the congruence of private investments in agriculture compared to the real needs and development potential of the countries receiving the investments.

It is to be noted that, due to the complexity and great inconsistency of the phenomenon of losses and waste, many studies have been carried out by international organizations and private and public establishments on the various aspects of the problem (UNEP et al., 2014; FAO, 2013b; WRI, 2013) in order to give useful tools and information to fight the phenomenon (study of causes, prevention campaigns, examples of best practices), both in developed and developing countries (IMechE, 2013; IMechE, 2014, HLPE, 2014).

However, it seems that the ongoing initiatives, though important, are destined to highlight their main limit: the inability to develop what, to us, seems to be the first and most effective premise to successfully contrast the problem of food losses and waste, i.e., the creation of a common conscience and common interests.

We have to consider that should, based on foreseeable demographic and economic evolutions, the current state of affairs go on, and should the problem of losses and waste not be tackled taking into consideration its global significance, the same is destined to grow, and with it the consequences that are already ensuing.

One of the main reasons for concern regarding the future of humanity is linked to the predictions made by FAO and OECD starting from 2009, which have presented the need to increase investments in agriculture and food production (FAO, 2009; OECD-FAO, 2015) to deal with the needs of a population estimated to grow to 9 billion by 2050 (+23.3% compared to today). The demand for raw materials for the production of biofuels and animal feed are also estimated to rise, since it is expected that these products will be in greater demand due to an income increase in many developing countries and, as far as biofuels are concerned, the European Union's choice to have, by 2020, 10% of the fuels market taken up by biofuels (Vieri, 1012c).

Notwithstanding the need to programme agricultural development on the basis of foreseeable demographic growth, it is clear that such predictions do not take into due account the fact that issues of sustainability, and therefore responsible use of resources available today, seem to be a prerequisite for availability itself, rather than issues regarding availability of resources.

We believe it is in this context that the problem of losses and waste should be contextualised. A significant reduction of what goes lost or wasted today could allow us not only to review the current predictions on the need to increase agricultural production and food, but also to handle the resources available today more responsibly, significantly cutting down the need to intensify or expand their use in the next future.

To achieve such an evolution it would be necessary, first of all, to radically rethink the models of development that have been put into practice up to today and which, through the current paradoxes of the agrifood systems, have shown their limitations. It should be clear that a problem such as that of losses and waste, being global, can be tackled efficiently only by involving as many actors as possible. This means that the actions against losses and waste will make a tangible difference only if they can favour the development of economic activities and individual behaviours more greatly linked to local resources and, therefore, more attentive to a responsible use of the same. In particular, and following the highlighted criticalities, it seems necessary to rethink the current relationship between economy, environment and society, building the prerequisites for an economic growth that identifies not only with the results made by large corporations operating in the global market in an oligopolistic regime, but also with the activities that generate widespread welfare, through inclusive models founded on the enhancement of the resources available in the region, or of single communities. Only thus a new mindset can be created that is able to contribute to the reduction of losses and waste with an approach that must be founded on the awareness of what is good and in the common interest and which, because of its nature, should be the main prerequisite to apply and support development models able to overcome the limits and criticality of the current ones.

Conclusion

The problem of losses and waste concerns about 30% of the world's agrifood production and is the main critical point of the current agribusiness development models. Although such phenomenon occurs with greatly different intensity and modalities based on the rate of socio-economic development of the single countries, it can be considered everywhere as the consequence of serious system inefficiencies. Because of this, the problem of food losses and waste must not be considered as a separate phenomenon, rather one in relation to the general context where it occurs. To this end it is important to identify both the main causes of the system's inefficiencies from which losses and waste derive, and also how they weigh on the main criticalities of the current agribusiness systems.

An important factor to determine said causes is to identify the distortions to the competition regime resulting from the physiological evolution of the rate of economic organization of the various parts of the agrifood supply chains which has lead to the creation of oligopolistic positions in the industrial and distribution sectors, upstream and downstream the agricultural production step.

Very important, as far as the interactions with other system criticalities are concerned, is the bearing of waste both for problems linked to malnutrition (hunger and excessive food consumption), and to issues of environmental sustainability (use of resources such as land and water, the use of chemical products, greenhouse gas emissions).

The seriousness and complexity of the problem of losses and waste needs an integrated and multi sectoral approach, to be studied for each single country, in the frame of a global strategy that should primarily be oriented to remove the causes, and therefore system inefficiencies, rather than act on the context that generates them. This is a further reason why we believe that, to efficiently contrast the problem of losses and waste, it is necessary, first of all, to develop a new mindset, to create new development models. Among the many famous expressions said to be uttered by Albert Einstein, one seems particularly apt, as a warning, in this context: "No problem can be solved from the same level of consciousness that created it."

REFERENCES

Anderson, K. (1987). On why agriculture declines with economic growth. *Agr Econ, 1(3),* pp. 195–207.

AQUASTAT. (2016). FAO's global water information system. http://www.fao.org/nr/aquastat.

Barilla Center for Food and Nutrition (BCFN). (2012). Food waste: causes, impacts and proposals. http://www.barillacfn.com/wp-content/uploads/2012/11/WEB_ENG.pdf.

Buzby, J. C., Hodan F. Wells and Jeffrey H. (2014). The Estimated Amount, Value, and Calories of Postharvest Food Losses at the Retail and Consumer Levels in the United States. Department of Agriculture,

Economic Research Service. www.ers.usda.gov/media/1282296/eib121.pdf.

FAO. (1981). Food loss prevention in perishable crops. www.fao.org/docrep/s8620e/s8620e00.htm.

FAO. (2009). *How to Feed the World.* Roma: FAO.

FAO. (2011). Global food losses and food waste – Extent, causes and prevention. www.fao.org/docrep/014/mb060e/mb060e.pdf.

FAO. (2012a). SAVE FOOD: Global Initiative on Food Loss and Waste Reduction. www.fao.org/save-food/en/.

FAO. (2012b). *Energy-Smart Food at FAO: An overview.* Rome: FAO.

FAO. (2013a). Food Wastage Footprint. Impacts on Natural Resources. www.fao.org/docrep/018/i3347e/i3347e.pdf.

FAO. (2013b). Toolkit: Reducing the Food Wastage Footprint, and related reports. http://www.fao.org/3/a-i3342e.pdf.

FAO. (2015a). *World Fertilizer Trends and Outlook to 2018.* Rome: FAO.

FAO. (2015b). Agroecology to Reverse Soil Degradation and Achieve Food Security. http://www.fao.org/3/a-i4803e.pdf.

FAO. (2016). Food Security Indicators. http://www.fao.org/economic/ess/ess-fs/ess-fadata/en/#.VlcBZNIvfGg.

Hanssen, O.J. and Møller, H. (2013). Food Wastage in Norway 2013. Status and trends 2013. http://www.fao.org.

HLPE. (2014). Food losses and waste in the context of sustainable food systems. A report by the High Level Panel of Experts on Food Security and Nutrition of the Committee on World Food Security, Rome 2014. http://www.un.org/en/zerohunger/pdfs/HLPE_FLW_Report-8_EN.pdf.

IMechE (Institution of Mechanical Engineers). (2013). Global Food-Waste Not, Want Not. http://www.imeche.org/knowledge/themes/environment/global-food.

IMechE. (2014). A Tank of Cold: Cleantech Leapfrog to a more food secure world. http://www.imeche.org/docs/default-source/reports/a-tank-of-cold-cleantech-leapfrog-to-a-more-food-secure-world.pdf?sfvrsn=0.

Lipinski, B., Hanson, C., Lomax, J., Kitinoja, L., Waite, R. and Searchinger, T. (2013). Reducing food loss and waste. World Resources Institute Working Paper, June. http://unep.org/wed/docs/WRI-UNEP-Reducing-Food-Loss-and-Waste.pdf.

OECD, and FAO. (2015). *OECD-FAO Agricultural Outlook 2015.* Paris: OECD Publishing.

Phillips McDougall. (2014). Directions in Global Research and Development for Crop Protection Products. http://apvma.gov.au/sites/default/files/docs/matthew_phillips_presentation_web-version.pdf.

Pimentel, D., Allen, J., Beers, A., Guinand, L., Hawkins, A., Linder, R., Stoner, K. (1993). Soil erosion and agricultural productivity. *World soil erosion and conservation*, pp. 277-292.

Ruttan, V. W. (1984). Models of Agricultural Development. In The Johns Hopkins University Press (ed), *Agricultural Development in the Third World* (pp 59-74). Baltimore: USA.

Rutten, M., Nowicki, P., Bogaardt, M.J. and Aramyan, L. (2013). Reducing food waste by households and in retail in the EU. http://edepot.wur.nl/290135.

Sen, A. (1981). *Poverty and famines: an essay on entitlement and deprivation.* Oxford: Oxford University Press.

Stop Wasting Food (SWF). (2013). TNS Gallup survey for Stop Wasting food movement and Danish Agriculture and Food Council. http://www.stopspildafmad.dk/inenglish.html.

UNEP, FAO, WRAP. (2014). Prevention and Reduction of Food and Drink Waste in Businesses and Households. http://thinkeatsave.org/index.php/be-informed/think-eat-save-guidance-document.

USDA. (2014). The Estimated Amount, Value, and Calories of Postharvest Food Losses at the Retail and Consumer Levels in the United States. http://www.ers.usda.gov/publications/eib-economic-information-bulletin/eib121.aspx.

Vieri S. and Calabrò G. (2014a). The impact of GMOs on the evolution of agribusiness systems. *Quality - Access to Success, Vol. 15, No. 142*, pp. 101-105.

Vieri S. and Calabrò G. (2014b). The use of GMOs and consumers' rights in the European Union. *International Journal of Environment and Health, Vol. 7, No. 2*, pp. 128-143.

Vieri S., (2012a). The common agricultural policy and measures for the protection and conservation of the environment: contrasts and balances and new development models for the future. *IJ Env h, Inderscience 6(1)*, 155-169.

Vieri, S. (2012b). *Agricoltura Settore Multifunzionale allo Sviluppo*. Bologna: Edagricole-IlSole24Ore.

Vieri, S. (2012c). Biofuels and EU's choices. *International Journal of Environment and Health, 6 (2)*, pp. 155-169.

WRAP. (2011). Reducing food waste through community focussed initiatives. http://www.wrap.org.uk/sites/files/wrap/2011.11_Worcestershire_CC_LFHW_2011_case_study.3e14035c.11397.pdf.

WRAP. (2013). Estimates of waste in the food and drink supply chain. http://www.wrap.org.uk/content/estimates-waste-food-and-drink-supply-chain.

WRAP. (2013b). Waste Prevention Case Studies. http://www. wrap.org.uk/content/waste-prevention-case-studies.

WRAP. (2015). Strategies to achieve economic and environmental gains by reducing food waste. http://newclimateeconomy.report/wp-content/uploads/2015/02/WRAP-NCE_Economicenvironmental-gains-food-waste.pdf.

WRI (World Resources Institute). (2013). Creating a Sustainable Future. http://www.worldresourcesreport.org/.

Biographical Sketch

Name: Simone Vieri

Affiliation: Department of Management - La Sapienza University of Rome, Italy

Email: simone.vieri@uniroma1.it

Education: Doctor in Agrarian Science

Address: Via del Castro Laurenziano 9- 00161 Rome (Italy)

Research and Professional Experience: Simone Vieri is full Professor of "Agricultural Economy and Policies." Since 1994 he has held various institutional offices, such as President of the National Institute of Agrarian Economy (2002–2006). He is author of many scientific monographs and articles on agricultural economics and policy; rapporteur at many seminars and conventions, national and international.

He has been head of the staff of many research projects in the field of agricultural economics and policy. His research activity mainly encompasses the following areas: common agricultural policy; national and regional agricultural policies, use of GMOs in agriculture and food; multifunctional agriculture; rural development.

Honored:

International Award "Verde Ambiente" under the patronage of UNEP, European Commission, Presidency of Italian Republic (2014)

Publications Last 3 Years:

Vieri S. (2013), Quality products and genetically modified organisms in Italy: hazards and possible enhancements, in Journal of Nutritional Ecology and Food Research, Vol. 1, 68-77.

Vieri S. (2013), Bello e di qualità: il contributo dell'agricoltura al futuro sviluppo della Nazione, in Verde e ambiente 2, 4-8.

Vieri S. (2013), I sistemi agroindustriali e le loro disfunzioni. Nel Mondo ne soffre una persona su quattro, Verde e Ambiente 4, 20-23.

Vieri S., Vinci G., Botrè F. Preti R., Tieri A. (2013), *GMO fate from the fodder to the food: the state of the art*, in Technolgy and Innovation for a Sustainable Future: A Commodity Science Perspective ISBN – 9788882862695.

Vieri S. (2013), Le misure di politica agraria in favore della difesa del suolo, in La meccanizzazione dell'agricoltura italiana. Aspetti tecnici, economici, ambientali e sociali, pp. 39-48, INEA, Roma.

Vieri S. (2013), Gli interventi pubblici a sostegno della meccanizzazione, in La meccanizzazione dell'agricoltura italiana. Aspetti tecnici, economici, ambientali e sociali, pp. 101-124, INEA, Roma.

Vieri S. (2013), Le macchine e l'agricoltura italiana, in La meccanizzazione dell'agricoltura italiana. Aspetti tecnici, economici, ambientali e sociali, pp. 125-153, INEA, Roma.

Vieri S., Mari F. (2013), Ipotesi metodologica per la valutazione dell'impatto economico e ambientale delle lavorazioni meccaniche in agricoltura, in La meccanizzazione dell'agricoltura italiana. Aspetti tecnici, economici, ambientali e sociali, pp. 83-100, INEA, Roma.

Vieri, S. (2014), Diritto ad un'alimentazione adeguata e modelli di sviluppo agricoli, in Verde Ambiente, 2, 6-11.

Vieri S., Calabrò G. (2014), The use of GMOs and consumers' rights in the European Union, in International Journal of Environment and Health, Vol 7, n. 2, 2014.

Vieri S, Calabrò G. (2014), The impact of GMOs on the evolution of agribusiness systems, Quality-Access to Success Journal, Vol. 15, n. 142, October 2014.

Vieri S, Calabrò G. (2014), The environmental certification of tourism: a tool to enhance the unicity of a territory, Quality-Access to Success Journal, November 2014.

Vieri S., Vinci G., Preti R. (2014), *Quality and origin assessment of extra virgin olive oil for consumer protection by chromatographic techniques*, in Innovazione, Sostenibilità e Tutela dei Consumatori: L'Evoluzione delle Scienze Merceologiche per la Creazione di Valore e Competitività pp. 571-582, Università di Pisa, Pisa.

Vieri S., Calabrò G. (2015), Beauty Will Save Us Hypothesis of European Structural Funds Use as a Tool to Create in Italy an Economic Model able to Create Widespread Wellbeing through the Exploitation of Beauty and Quality, International Journal of Economic Practices and Theories, Vol. 5, No. 3, 2015 pag.200-206.

Vieri S., Calabrò G. (2015), The strategic role of agriculture in a development system of exploitation of territorial resources, Proceedings BASIQ 2015 International Conference- New trends in sustainable business and consumption- Bucharest, 18-19 June 2015, pag. 164-170. ISSN/ISBN 2457-483X -ISSN-L 2457-483X.

Vieri S., Calabrò G. (2015), *C'est la beauté qui nous sauvera hypothèse d'utilisation des fonds socio-structuraux européens comme moyen pour la réalisation, en italie, d'un nouveaux modèle de développement économique*, in Atti della Conférence Internationale Compétitivité et cohésion économique et sociale, Bucarest, 26-27 Marzo 2015 p 17-22.

Name: Grazia Calabrò

Affiliation: Department of Economic - University of Messina, Italy
Email: grazia.calabro@unime.it
Education: Doctor in Economic
Address: Piazza Pugliatti 1, 98124 Messina (Italy)

Research and Professional Experience: Grazia Calabrò is Associated Professor of Commodity Science. She took part in a number of different working groups as principal investigator and co-investigator in the field of Commodity Science. Her scientific and research activity mainly encompasses the following areas: Quality Certification; Environmental Management and Certification; Green public Procurement; Tourism and Environment; Waste Management; Food Quality and Safety; Counterfeiting.

She is author of about 97 scientific publications including four monographs and has held several national and international conferences.

Honored: Young Research Award University of Messina, Italy (2004)

Publications Last 3 Years:

Vieri S., Calabrò G. (2015), Beauty Will Save Us Hypothesis of European Structural Funds Use as a Tool to Create in Italy an Economic Model able to Create Widespread Wellbeing through the Exploitation of Beauty and Quality, International Journal of Economic Practices and Theories, Vol. 5, No. 3, 2015 pag.200-206.

Vieri S., Calabrò G. (2015), The strategic role of agriculture in a development system of exploitation of territorial resources, Proceedings BASIQ 2015 International Conference- New trends in sustainable business and consumption- Bucharest, 18-19 June 2015, pag. 164-170. ISSN/ISBN 2457-483X -ISSN-L 2457-483X.

Vieri S., Calabrò G. (2015), *C'est la beauté qui nous sauvera hypothèse d'utilisation des fonds socio-structuraux européens comme moyen pour la réalisation, en italie, d'un nouveaux modèle de développement économique*, in Atti della Conférence Internationale Compétitivité et cohésion économique et sociale, Bucarest, 26-27 Marzo 2015 p 17-22.

Vieri S., Calabrò G. (2014), in International Journal of Environment and Health, ISSN online:1743-4963, ISSN print: 1743-4955.

Calabrò G., Vieri S. (2014) The environmental certification of tourism: A tool to enhance the unicity of a territory in CALITATEA-ACCES LA SUCCES, Vol. 15 November, ISSN 1582-2559, p. 44-54.

Popescu D., Saseanu A., Bulin D., Calabro G (2014) *Econometric models in romanian tourism under the impact of sustainable development* in AMFITEATRU ECONOMIC, ISSN 1582-9146, pag 1063-1075.

Simone Vieri S., Calabrò G. (2014) *The impact of GMOs on the evolution of agribusiness systems in* CALITATEA-ACCES LA SUCCES, ISSN 1582-2559 vol 15 (142) p 101-105.

Calabrò G. (2014), Economia Low Carbon: Validità Della Scelta Dei Biocarburanti Di Origine Agricola, in Atti del XXVI Congresso Nazionale.

di Scienze Merceologiche Innovazione, Sostenibilità e Tutela dei Consumatori: L'Evoluzione delle Scienze Merceologiche per la Creazione.

di Valore e Competitività, a cura di A. Tarabella, pp 132-143, ISBN 978-1-291-74318-0.

Calabrò G. Marketing strategies for the agri-food sector: adaptation requirements and contradictions in Salomone R., Saija G. (2014) Pathways to environmental sustainability: methodologies and experiences. Springer International Publishing AG, Cham, Dordrecht, The Netherlands, ISBN 978-3-319-03825-4, p. 43-52.

Calabrò G., Lagioia G. (2013) L'etichetta climatica sui prodotti alimentari: un nuovo approccio dell'industria alimentare- Industrie Alimentari, ANNO 52- N. 534 Aprile 2013- Chiriotti Editore- ISSN 0019-901x; pag. 15-21.

Reviewed by: Stefano Masini- Department of medicine and surgery - "Tor Vergata" University of Rome–- Via Montpellier, 1- 00133 Rome- E-mail: stefano.masini@uniroma2.it

In: Food Waste
Editor: Garrett Leonard Riley
ISBN: 978-1-63485-025-4

Chapter 2

FOOD WASTES AS SOURCES OF NATURAL ANTIOXIDANTS

Adriana S. Franca and Leandro S. Oliveira
Universidade Federal de Minas Gerais
Belo Horizonte, MG, Brasil

ABSTRACT

Antioxidants are key ingredients in functional foods, given their well established health benefits. They act by neutralizing free radicals, i.e., substances that occur naturally in our bodies but attack fats, protein and the DNA in our cells, either causing different types of diseases or accelerating the aging process. Synthetic antioxidants (e.g., TBHQ, BHA and BHT) have been widely used as food additives, but their application has been currently reassessed because of possible toxic or carcinogenic components formed during their degradation. Consequently, the search for endogenous protective ingredients in foods has been intensified and a wide variety of natural antioxidants have been added to food products, mainly polyphenols. These compounds are secondary metabolites of plants, and are generally involved in defense against ultraviolet radiation or aggression by pathogens. Phenolic compounds have been shown to exhibit a wide range of physiological properties, including anti-allergenic, anti-atherogenic, anti-inflammatory, anti-microbial, antioxidant, anti-thermobiotic, cardio protective and vasodilatory effects. As food ingredients, they have elongated the shelf life and oxidative stability of stored products and also increased their potential as functional foods. An extensive amount of plant biomass wastes are produced yearly as by-products from the agricultural and food industries. Given the protective action of polyphenols in plants, higher concentrations of such compounds are present in peels, skins and seeds, thus supporting the use of these residues as a source of natural antioxidants. In view of the aforementioned, the objective of the present study is to present a review of the works of research that have been developed in the last decade on the use of food wastes as sources of antioxidant compounds, such as polyphenols.

1. Introduction

An antioxidant is any substance that, in concentrations lower than those of an oxidizable substrate, significantly delays or prevents oxidation of the substrate (Halliwell and Gutteridge 2007). Hence, antioxidants constitute a relevant class of additives with the ability to protect food against detrimental changes due to undesirable oxidation of nutrients (André et al. 2010), and the ability to minimize the adverse effects of oxygen and nitrogen reactive species (e.g., free radicals) in biological systems such as the human body (Mishra et al., 2012). As food additives, antioxidants hinder oxidative rancidification, prevent formation of toxic oxidation products and extend shelf life of the food products. Lipid oxidative rancidification is one of the major causes of food quality deterioration and addition of antioxidants has been deemed the most effective method to control its rate (Shahidi and Ambigaipalan, 2015). In biological systems, antioxidants prevent free radical induced tissue damage by preventing the formation of radicals, by scavenging them, or by promoting their decomposition (Young and Woodside, 2001).

Antioxidants are classified as primary, which are active radical scavengers or chain reaction breakers, e.g., polyphenols and secondary aryl amines; and secondary, which are peroxide decomposers, e.g., organophosphites and thioesters (André et al., 2010). Polyphenolic compounds are ubiquitous in the plant kingdom and are the most abundant secondary metabolites of plants, being present in structural arrangements that range from simple molecules to high molecular-weight polymers, consisting mostly of derivatives and/or isomers of flavones, flavanols, flavanones, isoflavones, anthocyanins and phenolic acids. These compounds are an essential part of the human diet and their consumption has been associated to lower incidence of human degenerative diseases, such as cancer, and to decreased risk factors of cardiovascular diseases (Kris-Etherton et al., 2002), and these beneficial health effects are mostly attributed to their antioxidant activity (Shahidi and Ambigaipalan, 2015). Thus, in view of the growing evidence of their potential health benefits, dietary polyphenols are amongst the most investigated compounds over the past decades as ingredients in functional foods and beverages (Lee, 2013). The major sources of these dietary phenolic compounds are fruits, vegetables, cereals, coffee, chocolate, wine, and many edible plants.

In light of the growing awareness of the population regarding the effects of healthier diets on the quality of living, society has undergone changes in eating habits, and these changes resulted in increased demand for dietary food products, with a consequent increased production of these products over the past decades (Gómez et al., 2015). Aside from the benefits of adequately feeding larger portions of the population, a major drawback of increasing food production is the concurrently increase in the amount of waste generated. The agricultural practices alone generate wastes that range from 50 to 300% of the targeted product weight (Haq and Easterly, 2006; Purohit et al., 2006; Lal, 2008; Lora and Andrade, 2009) and the fruit and vegetable industries are estimated to generate wastes in the range of 5 to 50% of the raw material weight (Gómez et al., 2015). Food waste generation produces significant negative impacts at environmental, social and economic levels. From an environmental point of view, agricultural and food wastes contribute to Green House Gas emissions when they are disposed in landfills (via uncontrolled methane release) and to volatile organic compounds emissions when they are burnt for energy recovery. Other environmental impacts associated with these wastes are natural resource depletion in terms of soil, nutrients, water and energy, and disruption of biogenic cycles due to intensive agricultural activities. Social impacts may be ascribed to ethical and moral dimension within the general concept of global food security, and economical impacts are related to the costs of adequately dealing with wastes and their effects on producers and consumer incomes (Girotto et al., 2015). Adequate disposal of agricultural and food wastes is a major issue since biological materials are usually prone to fast microbial spoilage, limiting further exploitation. Further, if discarded to soils and water bodies, these wastes present a high biochemical oxygen demand for its proper degradation, and may lead to undesirable eutrophication of the local ecosystem and, hence, the problem of waste disposal is further aggravated by legal restrictions. Therefore, efficient, inexpensive and environmentally sound exploitation of these materials have become rather relevant in the past years.

Of all the options being investigated and proposed for minimization of by-products (waste) in the agricultural and food sectors, recovery of bioactive compounds is one of the most promising options. These types of waste constitute cheap sources of biologically active chemicals (e.g., polyphenols) and of functional food ingredients (e.g., dietary fibers), and existing technologies allow for their recovery and further use as functional ingredients in food products (Galanakis, 2012). Fruits and vegetables processing wastes are mostly comprised of soft tissues rich in antioxidant compounds and dietary fibers, and consequently are the most widely investigated substrates for the profitable recovery of these components. A significant body of research work has been published in the past years regarding developing methodologies for the quantification and identification of bioactive compounds in agricultural and food wastes, and also on methodologies for their profitable recoveries, as reviewed by Schieber et al. (2001), Laufenberg et al. (2003), Santana-Méridas et al. (2012), Galanakis (2012, 2013), Girotto et al. (2015) and Gómez et al. (2015).

In view of the aforementioned, it is the aim of this essay to present an overview of recent studies that deal with the use of food wastes as sources of antioxidants. It should be emphasized at this point that the literature pertaining the study of recovery of polyphenols and other antioxidant compounds from agricultural and food wastes is quite abundant and it is not the purpose of this chapter to cover the extensive information they present, but rather highlight and critically analyze current works on this area that set the trends for future research and applications.

2. Recovery of Antioxidant Compounds from Food Wastes

2.1. Recovery of Antioxidants for Application as Functional Food Ingredients

Large amounts of water are employed in the continuous washing of the olive paste during olive oil production and together with that endogenously contained in the olives they will comprise the olive mill wastewater (OMWW). Olive mill wastewaters are discarded by the olive oil manufacturers, representing a significant burden to the mills' economy. A research work carried out by Visioli et al. (1999) has demonstrated the

OMWW to contain a considerable amount of phenolic compounds associated with potent antioxidant properties. Thus, with the objective of upgrading the byproducts of the olive oil industry, an investigation was carried out by Visioli et al. (1999) to evaluate different procedures for the recovery of active components of OMWW and to compare their antioxidant activities. OMWW were demonstrated to be rich in recoverable antioxidant compounds, and OMWW extracts were able to inhibit human LDL oxidation and scavenge superoxide anions at low concentrations. Since the late 1990s, olive mill wastes and wastewaters have been receiving much attention and OMW management have covered all possible strategies towards a biorefinery concept: treatment, disposal, recovery and reuse. In view of this scenario, Federici et al. (2009) presented a thorough review of the findings up to 2009 with a focus on antioxidants recovery and waste biotechnological valorization in the fermentative production of enzymes of actual or potential commercial interest. According to Federici et al. (2009), olive mill wastewaters (OMW) are characterized by large volumes and high polluting load, with a high content of compounds with biostatic activity (e.g., polyphenols). Several disposal approaches have been proposed for OMW, and, in some cases, applied to this waste. Examples of such approaches are decantation with lime, concentration, drying and incineration, ultrafiltration, reverse osmosis, aquaculture, land-spreading, animal-breeding, biological treatment in aerated lagoons, and anaerobic digestion. OMW is rather rich in polyphenols with reasonable contents of low-molecular weight phenolic compounds, such as benzoic acid derivatives and hydroxycinnamic acid derivatives, and with high concentrations of tyrosol, homovanillyl alcohol and hydroxytyrosol. These compounds are of great interest for they exhibit high antioxidant activities and, after filtration to eliminate the suspended solids, can be recovered by physico-chemical processes such as ultrafiltration, nanofiltration and reverse osmosis. Hydroxytyrosol, the main polyphenol in OMW, possesses the most potent antioxidant activity, similar to those of synthetic antioxidants, such as BHT (2,6–ditert-butyl-phydroxytoluene). Hydroxytyrosol was obtained by total synthesis and commercialized at very high market price and its presence in significant concentrations in OMW has turned it of great importance for direct recovery (Allouche et al., 2004; Gortzi et al., 2008). By using the surfactant Genapol X080, Gortzi et al. (2008) were able to attain enhanced recovery of this polyphenol in various concentrations. Federici et al. (2009) stated that, of even greater commercial relevance, was the development of environmentally friendly and economically sustainable chemical procedures to convert the compounds recovered from OMW into hydroxytyrosol, citing the works by

Bernini et al. (2008) in which hydroxytyrosol and its lipophilic derivatives were synthesized from tyrosol or homovanillyl alcohol.

Galanakis et al. (2010) have also worked with OMW in a study to investigate the effect of ethanolic extraction and pre-treatment parameters on the phenolic content and antioxidant capacity of the resulting extracts with the goal to optimize phenol recovery from olive mill wastewater. Extractions were carried out at the time intervals of 1, 30, 60 and 120 min and the phenolic content, in total as well as in different classes, were determined for each time interval, together with the antioxidant capacity. These characteristics were monitored over a period of 18 weeks in order to evaluate the extracts stability and increase the possibility of further utilization. The total phenolic content and hydroxycinnamic acids concentrations of the extracts recovered after 30, 60 and 120 min extraction ranged from 676 to 691 mg L^{-1} and from 163 to 165 mg L^{-1}, respectively, and the extract recovered at 1 min exhibited significantly lower values than those for the longer extraction times. No significant differences were observed for antioxidant capacity between the four tested extracts. Heating of OMW had a negative impact on the recovered extracts incurring in reduction of the phenolic contents and antioxidant capacity of the extracts. Nevertheless, the extracts were able to retain the contained phenols during storage for 18 weeks. Phenol concentrations of the extracts were appropriately preserved in a solution of 85 mL ethanol per 100 mL water as well as the corresponding antioxidant activity. In some cases, the antioxidant activities of the extracts were improved during storage for 18 weeks.

Cardinali et al. (2010) investigated the antioxidant activities of OMW high molecular weight phenolic compounds fraction and extracts by a group of bioassays with different molecular targets. The OMW sample was separated into a clear sample (untreated OMW), an ultrafiltrate sample containing 60% of the total phenolics and comprising of monomeric and oligomeric phenolics (MW < 3000 Da), and a sample containing the remaining 40% of the total phenolic content, comprised of polymeric phenolics (MW > 3000 Da). Both the ultrafiltrate fraction and the polymeric fraction exhibited high antioxidant activities in the control of oxidative damage caused by peroxyl or hydroxyl radicals, with this activity being attributed to the complex phenolic content of the samples. The results for the HPLC analysis demonstrated the clear sample to contain the monomeric phenolics tyrosol (0.223 g/L) and hydroxytyrosol (1.112 g/L), with these two compounds representing about 30% of the total phenolic content of the ultrafiltrate fraction and about 20% of the total phenolic content of the clear OMW. Both these monomeric phenolics were absent in the polymeric fraction. Since both ultrafiltrate and polymeric fractions exhibited high antioxidant activities, the authors concluded that the antioxidant activity of OMW could not be completely attributable to monomeric phenolics and decided to recover oligomeric phenolics from clear OMW, by means of membrane filtration technology, and to evaluate their biological activity. The recovered fraction contained phenolic compounds with a molecular weight in the range of 200-5000 Da (about 52% of total phenolic content of clear OMW) and HPLC-MS analyses revealed the presence of verbascoside and isoverbascoside in this ultrafiltrate fraction, which also contained tyrosol and hydroxytyrosol. The ultrafiltrate fraction was further submitted to chromatographic separation on Sephadex LH-20 to yield an OMW extract free of tyrosol and hydroxytyrosol (LH-20 fraction). This fraction was analyzed by HPLC-MS/MS, revealing the presence of phenolic compounds with molecular masses in the range of 600 to 1000 Da. Besides verbascoside and isoverbascoside, the analysis of the LH-20 fraction has also revealed the presence of an oxidized form of verbascoside and a number of higher molecular weight phenolics arising from oxidative polymerization of hydroxytyrosol and caffeic acid. In order to evaluate possible protective effects of cells against an induced oxidative stress (H_2O_2) using flow cytometry, OMW extracts and standard hydroxytyrosol were assayed on HT29 human colon carcinoma cells by using a DCFDA probe. Concentration-dependent antioxidant activities were observed for both the ultrafiltrate (MW $<$ 5000 Da) and the LH-20 (MW = 600-1000 Da) fractions. It was also observed that high molecular weight phenolics exhibited a capacity to inhibit oxidative stress greater than those of low molecular weight phenolics. Thus, it was concluded that, aside from targeting the maximum recovery of hydroxytyrosol, recovery of high molecular weight phenolics from could be regarded as a source of natural antioxidants.

OMW was also investigated by Lafka et al. (2011) to determine its phenolic content and antioxidant capacity. OMW was subjected to conventional liquid solvent extraction, using different solvents (methanol, ethanol, mixture of ethanol to water 1:1, n-propanol, isopropanol and ethyl acetate), and to supercritical fluid extraction (SFE), using carbon dioxide. The optimum solvent for the extraction of phenolics was ethanol and the optimum extraction conditions were 180 min, a solvent to sample ratio of 5:1 v/w, and pH 2. The predominant phenolic compound in the HPLC-analyzed OMW was hydroxytyrosol, and phenolic acids and flavonoids were also identified The extracts obtained by solvent and supercritical fluid extractions were tested for their antioxidant activity with the ethanol extract exhibiting the highest antiradical activity (DPPH·). No correlation was observed between antiradical activity and phenolic contents. Although the SFE extract presented a low phenolic content, it exhibited a reasonable antioxidant capacity. The ethanol extract was deemed a stronger antioxidant than BHT, ascorbyl palmitate and vitamin E by the Rancimat method on sunflower oil.

Angelino et al. (2011) isolated a phytocomplex from olive mill waste waters, comprised of the phenolic compounds hydroxytyrosol, tyrosol and the dialdehydic form of decarboxymethyl elenolic acid, linked with (3,4–dihydroxyphenyl)ethanol, which accounted for 6% of the total phenolic content. The phytocomplex was evaluated for antioxidant capacity measured by the cellular antioxidant activity in red blood cell method, and it was demonstrated that the phytocomplex was able to permeate the cell membrane, exhibiting antioxidant activity inside the red blood cells. A polyphenolic fraction of OMW was obtained by Cardinali et al. (2012) using membrane process and chromatographic purification on Sephadex LH-20, and verbascoside was identified as the most abundant phenolic compound in the LH-20 fraction. This fraction was further purified to obtain a verbascoside with a high degree of purity and the inhibitory effects of verbascoside against oxidative stress in cultures of HT-29 human colon carcinoma cells as well as its capacity to inhibit LDL oxidative modification induced by metal catalysts was investigated. In the studied HT-29 cell system, verbascoside was incorporated in a concentration- and time-dependent manner, with an accumulation efficiency of about 0.12%, and significantly counteracted the cytotoxicity of reactive oxygen species (ROS) generated from H_2O_2 prior to attacking biomolecules. Its antioxidant activities in cell systems was partially attributed to its significant affinity for negatively charged phospholipid membranes. Verbascoside was also demonstrated able to protect LDL against copper catalyzed oxidation and this capacity was attributed to its ability as an electron donor, independently of its ability to chelate divalent copper. Considering all the antioxidant assays employed in the study, the antioxidant capacity of verbascoside was 2 to 10 times higher than that of the LH-20 fraction as a whole, which includes tyrosol and hydroxityrosol. Thus, verbascoside was deemed a potent antioxidant biophenol with promising applications in the cosmetics, nutraceutical and functional food industry.

Phenolic extracts from OMW were obtained by El-Abbassi et al. (2012) by microfiltration followed by liquid-liquid extraction with ethyl acetate, vacuum evaporation at 40°C to complete dryness and resolubilization in methanol. Olive mill wastewater samples were collected from two different olive oil mills employing different milling techniques. The phenolic composition of extracts was determined by HPLC and the extracts were tested for their antioxidant capacity using a diversity of assays. Hydroxytyrosol was the most abundant phenolic compound in the OMW extracts, comprising about 70% of the total phenolic concentration of the extracts from the mill that uses more water and 55% of the total phenolic concentration of the extracts from the mill that uses less water. Other phenolics were tyrosol, caffeic acid, gallic acid, p-coumaric acid, and oleuropein and its aglycone. Both the microfiltrate fraction and the ethyl acetate extract presented considerable antioxidant capacity, with the microfiltrate fraction presenting higher activity. This higher activity for the microfiltrate fraction was attributed to its diverse phenolic composition, which included heavier phenolics that were not present in the ethyl acetate extracts, since the ethyl acetate extraction is more selective for low and medium molecular weight phenols (Visioli et al., 1999).

Ena et al. (2012) evaluated the efficiency of dry Azolla caroliniana, a freshwater aquatic fern, and granulated activated carbon (GAC) in adsorbing and desorbing polyphenols as a process for their recovery from OMW. The powder products recovered by the desorbing processes were characterized for their phenolic contents and antioxidant activities. Desorption was carried out by acidified water at pH 3 using HCl for the Azolla adsorbent and by ethanol at pH 11 using NaOH for GAC. The polyphenol content of the powder obtained by desorption from the Azolla adsorbent was 2.13 times higher than the obtained by desorption from GAC, and this difference was attributed to the ability of the pectin present in Azolla to electrostatically interact with the positively charged polyphenol molecules at the adsorption solution pH. The phenolic compounds desorbed from Azolla were identified as hydroxytyrosol, tyrosol, and gallic, vanillic, caffeic, syringic and coumaric acids. The most abundant components of the Azolla desorption product were gallic acid and tyrosol. Only five biophenols were identified in the GAC desorption product: hydroxytyrosol, tyrosol, and gallic, syringic and coumaric acids. The concentration of hydroxytyrosol in the GAC desorption powder was higher than in the Azolla desorption product, demonstrating GAC to possess high affinity for this small molecule. The total polyphenol content in the Azolla product was about two times higher than in the GAC product. Nevertheless, the antioxidant activity of these products were quite similar, a fact attributed to the differences in polyphenol composition of the desorption products. Hydroxytyrosol is the most potent antioxidant among the polyphenols found in the desorption products and it was present in higher concentrations in the GAC product than in the Azolla product. The correlation between the polyphenol content and the antiradical activity of both Azolla and GAC desorption powders was found to be almost linear. Both Azolla and GAC were deemed suitable matrices for both adsorptive recovery of OMW phenolic compounds. Uribe et al. (2015) studied the olive-waste cakes, from the processing of the varieties Picual, Frantoio and Arbequina, for their phenolic, tocopherol, flavonoid, flavanol, dietary fiber and β-carotene contents, as well as for their antioxidant activity. The polyphenols tyrosol and hydroxytyrosol were detected in the free extracts of the three varieties. The Picual variety presented the highest free phenolic and free flavonoid contents and, accordingly, the highest antioxidant capacity, while the Arbequina variety presented the highest β-carotene content.

Red ginseng marc (RGM) is generated and discarded as waste in large amounts in processing plants due to an increased number of red ginseng products used in the functional food industry. Lee et al. (2009) have determined that red ginseng marc contained bioactive components and, thus, Lee et al. (2015) evaluated the bioactive components and properties of RGM extracts prepared via fermentation with *Bacillus subtilis* (BSRGM) and *Saccharomyces cerevisiae* (SCRGM). No polyphenols were detected in the non-fermented RGM (NRGM) extracts, whereas polyphenol contents in RGM extracts significantly increased following with fermentation. The polyphenol contents in BSRGM and SCRGM were respectively 7.226 and 6.875 g/100 g, and the fermented extracts exhibited significantly higher radical scavenging activities than those of NRGM extracts, suggesting a positive correlation between polyphenol content and antioxidant activity, as measured by free radical scavenging activity. However, although the non-fermented extracts did not contain polyphenols, they exhibited radical scavenging activities and the authors inferred that these activities were due to the presence of other bioactive compounds, such as uronic and sialic acids.

Onion solid wastes (OSW) are comprised of the non-edible part of the onion bulb, i.e., the outer dry and semi-dry layers, and the apical trimmings, with an estimated 6,000,000 tons being generated in Europe on an annual basis (Katsampa et al., 2015). In previous investigations, with the employment of water/ethanol mixtures as solvents, Khiari et al. (2009) and Kiassos et al. (2009) demonstrated OSW extracts to be significantly richer in polyphenols than the extracts of the edible counterpart. It was also demonstrated by these authors that the polyphenol composition of OSW was different from the polyphenol composition of the onion edible part. Despite its non-toxicity and ability to extract polyphenols, ethanol presents a relatively high cost and its disposal is regulated by State laws, rendering its recovery and reuse a strict necessity. Thus, Katsampa et al. (2015) used an aqueous solution of glycerol as an inexpensive, abundant and non-toxic solvent in their proposed green ultrasound-assisted extraction process for the recovery of antioxidant polyphenols from onion solid wastes. The maximum yield of total polyphenols at saturation, at 50◦C, was 90.07 mg GAE g^{-1} dw, a level similar to that achieved by Kiassos et al. (2009) with 60% ethanol at pH 2 (93.42 mg GAE g^{-1} dw) and significantly higher than that achieved by Khiari et al. (2009) with 60% ethanol/0.1% HCl (20.83 mg GAE g^{-1} dw). With the ultrasound-assisted extraction, the yield of polyphenols at saturation, at 80◦C, was 1.14 times higher than that achieved with pH-regulated extraction of OSW with 60% ethanol (Makris, 2010). The reducing power, a measure of the antioxidant activity, was determined to increase by approximately 1.6 times, when the extraction temperature was increased from 50 to 80◦C. Thus, it was inferred that no loss of antioxidants occurred with the increase in the extraction temperature and that the higher yield of total polyphenols attained at 80◦C may be closely associated with higher antioxidant activity observed.

Tana et al. (2012) proposed a clean and economical way of producing a caffeine-free polyphenol-rich extract from green tea waste in one-step by Sephadex LH-20 column chromatography. Sephadex LH-20 was used for its unique chromatographic selectivity due to dual hydrophilic and lipophilic nature. In a stepwise manner, caffeine and other components were removed by elution with a 20% (v/v) ethanol solution and, sequentially, tea polyphenols were eluted with a 60% (v/v) ethanol solution by reversed flow. The yield of polyphenols was higher than 11%, with a purity of about 98%, and with catechins comprising more than 91% of the total polyphenols. Caffeine was not detected in the produced extract as demonstrated by high performance liquid chromatography. The recovery of tea polyphenols by column chromatography were about 86%, and the stability of preparative way is very good. Compared with conventional extraction and decaffeination processes, the proposed method of polyphenol-rich extract preparation was deemed advantageous, for it employs cheaper raw material, generates no toxic residues, leads to higher polyphenol yields, and is commercially feasible and environmentally favorable.

Wine pomace (WP) are the solids remaining after the fermentation of red grapes, racking-off the wine, and subsequent pressure, and are mainly comprised of solid grape skin, rest of pulp, seeds, and residual yeasts and bacteria, the agents that carry out the fermentations. As discussed by Yu and Ahmedna (2013), WP contains a plethora of bioactive phytochemicals and great potential to serve as a source of functional food ingredient. In light of this, García-Lomillo et al. (2014) evaluated the ability of wine pomace to prevent oxidative degradation of foods, thus extending their shelf life, and to control the growth of spoiler microorganisms. The waste materials were obtained and studied in three forms: Whole wine pomace product (WWPP); seed-free WP, termed skins wine pomace product (SkWPP); and isolated seeds, termed seeds wine pomace product (SdWPP). The three types of by-products exhibited significantly different chemical compositions. In relation to antioxidant activity, the total polyphenol content of SdWPP was higher than that of WWPP, which in turn was higher than that of SkWPP. As expected, the total anthocyanin content of SkWPP (mainly comprised of skin and rest of pulp) was higher than that of SdWPP (skin free), and the total proanthocianidin content of SdWPP was higher than that of SkWPP. All three materials were UV and thermally treated and the effects of treatment on their chemical composition were evaluated. The treated products exhibited proximate compositions similar to those of the untreated ones with differences in phenolic composition being observed. WWPP presented significant decreases in total polyphenol content for both types of treatment (16% in UV-treated and 6% in thermally treated WWPP). Total polyphenol content of SdWPP was significantly affected only by the UV treatment, with a reduction of about 10%, and the content of WWPP was decreased by both treatments. The decreases in total polyphenol content caused by the UV treatments were attributed to photodimerization and isomerization of phenolic compounds induced by UV radiation, and to possible formation of new linkages between hydroxycinnamic acids and lignin units that would decrease their extractability. The antioxidant activities of the wine by-products were evaluated by the ABTS method and by the Fat Oxidation Inhibition Capacity method, using the Rancimat test. Regarding the ABTS method, the three WP products presented statistically significant differences in their antioxidant capacities, with SdWPP exhibiting the highest value (141.99 ± 2.09 μmol/g), followed by WWPP (103.29 ± 0.23 μmol/g) and SkWPP (75.65 ± 1.98 μmol/g). ABTS results positively correlated with total polyphenol, total catechin and total proanthocianidin contents of the three by-products, suggesting the attribution of antioxidant activity, for the most part, to their phenolic composition. Regarding the fat oxidation inhibition capacity, the results were deemed to correlate well with the ABTS measured antioxidant capacities of the by-products and thus to their polyphenolic contents. Although SkWPP presented lower total polyphenol content than that of SdWPP, protection against fat oxidation exerted by SkWPP was significantly higher than that by SdWPP, and this higher protection was attributed to its higher content of anthocyanins, which have already been demonstrated to present an excellent ability to scavenge the free radicals formed during fat oxidation (García-Lomillo et al., 2014).

Silva and Jorge (2014) investigated the bioactive compounds (essential fatty acids, tocopherols, phytosterols, phenolic compounds and carotenoids) and antioxidant capacity of oils obtained by extraction from grape, guava, melon, passion fruit, pumpkin, soursop, and tomato disposed seeds. Tocopherol is the main antioxidant compound in oils and the total quantity of tocopherols in the oils ranged from 29.2 mg/kg in soursop to 328.7 mg/kg in tomato seed oils, respectively. The most significant isomer was γ-tocopherol, except in passion fruit seed oil, in which the major isomers were β + δ-tocopherol, and in soursop seed oil, which presented higher quantity of α-tocopherol. The main phenolic compounds determined in the oil samples were catechin in grape seed oil, p-coumaric acid in guava seed oil, salicylic acid in passion fruit seed oil, and epicatechin in soursop seed oil. Carotenoids, a group of pigments with lipid oxidation-prevention activity were detected in all oil samples, with total carotenoids content ranging from 4.4 μg/g for soursop seed oil to 11.9 μg/g for grape seed oil. The antioxidant capacity of oils were determined by the methods DPPH•, ABTS•+, FRAP, and β-carotene/linoleic acid. Tomato seed oil presented the highest antioxidant capacities for the methods ABTS•+ and β- carotene/linoleic acid, and the lowest performance with the FRAP method. Soursop seed oil presented the highest radical scavenging activity for the DPPH• method.

The industrial processing of mango generates significant amounts of by-products, with 35 to 60% of the fruit being discarded as waste, which is mainly comprised of seeds, peels, and paste. These by-products are a valuable source of dietary fiber and bioactive compounds, such as polyphenols, and thus are potential candidates as ingredients for functional foods. Hence, Blancas-Benitez et al. (2015) prepared starch-molded mango snacks by adding dried 'Ataulfo' mango by-products as source of dietary fiber and polyphenols as a replacement for sugarcane. The peel and paste of dried 'Ataulfo' mango by-product (DAMB) were added to the mango concentrate (DAMC) at different ratios of DAMC:DAMB and the physical and sensory properties, the dietary fiber and polyphenol contents, and antioxidant capacity were determined for the resultant starch-molded mango snacks. Regarding the control snack formulation (i.e., snack produced without addition of by-products), the total dietary fiber content increased about 2.5 times whereas polyphenol content increased about five times when sugarcane was replaced by 25:75 DAMC:DAMB. The mango by-product-enriched snack exhibited higher antioxidant activity than the control one. Incorporation of dried mango by-products to starch-molded mango snacks was demonstrated to improve the dietary fiber and polyphenol contents of mango snacks by reducing its sugar content with no significant effects on their physical-chemical properties. Sensory characteristics, however, were deemed in need of improvement.

Blidi et al. (2015) comparatively evaluated the performance of three bio-solvents for the extraction of polyphenols from apple waste peels. The studied solvents were aqueous solutions of ethanol, butanediol and glycerol. All the studied solvents performed well, with the extract obtained with 50% (v/v) ethanol exhibiting higher reducing power, i.e., higher antioxidant activity than the other extracts. Liquid chromatography-mass spectrometry analysis demonstrated the extracts to contain twelve major polyphenols. However, differentiated selectivity was presented by the solvent glycerol (70% v/v) which produced an extract with the absence of four of the p-coumarate conjugates. The authors concluded that the three bio-solvents evaluated present great potential for the development of eco-friendly industrial processes for the production of polyphenol-rich products and that the selection of either solvent would not compromise extraction yield.

Ambigaipalan and Shahidi (2015) investigated antioxidant activity as well as solubility and water-holding capacity of date-seed protein hydrolysates in various food (cooked comminuted fish model and β-carotene–linoleate model) and biological (human LDL and DNA) model systems. Date seed proteins were hydrolyzed and the hydrolyzate compared to carnosine, an antioxidative peptide, with respect to functional and antioxidative properties. The incorporation of hydrolysates in fish model systems resulted in higher inhibition of oxidation then the synthetic antioxidant butylated hydroxytoluene. In addition, the hydrolysates inhibited β-carotene and LDL cholesterol oxidation, and inhibited hydroxyl and peroxyl radical-induced DNA scission. In view of the results, date seed protein hydrolysates were deemed a potential functional food ingredient for health promotion.

The physicochemical and microbiological properties and the bioactive compounds were determined by Soquetta et al. (2016) for flours made from the skin and bagasse of two varieties (Bruno and Monty) of kiwi fruit (A. deliciosa) at two stages of maturation. Significant differences were observed regarding the phenolic compounds in the kiwi flours for all the treatments, with the flours made from the skin presenting higher contents than the flour made from bagasse. For both varieties, the flours made from kiwi fruit skin exhibited higher content of bioactive compounds and antioxidant activity than the flour made from kiwi fruit bagasse. The flours from the Monty variety of kiwi fruit exhibited the highest contents of flavonoids, chlorophyll, carotenoids and vitamin C, and also the highest antioxidant activity using the FRAP method. The flours from the Bruno variety of kiwi fruit exhibited the highest contents of phenolic compounds and the highest antioxidant activity using the DPPH• method.

Mustafa et al. (2016) investigated the potential of the by-product of a bitter-liqueur production as a source of bioactive compounds, in order to avoid the high costs associated with its disposal. The liqueur was comprised of a mixture of herbs, roots, barks, flowers and fruits such as gentian, rhubarb, cinchona, cinnamon, clove, star anise and orange with added sugar or other sweeteners. In its production process, the mixture was steeped in food-grade alcohol and matured to yield the targeted alcoholic beverage, with an alcoholic strength of 34% v/v. The by-product precipitated during liqueur production was collected and analyzed for total content of polyphenolics and free radical scavenging activity of the extracts was determined by the DPPH• assay. The antioxidant activity of ethanol and ethanol:ethylacetate (1:1) extracts of the bitter-liqueur by-product using the DPPH• assay and its correlation with the total polyphenolic content was investigated. The total polyphenol content of the ethanol:ethylacetate extract was twice as that of the ethanol extract (127.66 and 256.55 mg of GAE/g of extract, respectively) and the antioxidant activity of the ethanol:ethylacetate extract was higher than that of the ethanol extract. Furthermore, the scavenging activity of the ethanol:ethylacetate extract was higher than that of the standard antioxidant BHT, indicating that this extract exhibits a strong antioxidant activity.

The saffron spice is produced from the stigmas of Crocus sativus L. flowers and, in its production chain, only the stigmas are collected and used as a spice, while the remaining parts of the flower (90% in weight) are discarded as waste. However, the discarded parts of the flower were demonstrated to present great potential for use in the health and food industries, due to their content of bioactive compounds, such as carotenoids, polyphenols, anthocyanins and flavonoids. These bioactive compounds are known to present high antioxidant activities. Thus, Tuberoso et al. (2016) evaluated the *in vitro* antioxidant activity of juices obtained from cold-pressing of saffron floral by-products and investigated the composition of the polar fraction of the extracts by means of high resolution mass spectrometry. Cold-pressed juices obtained 24 and 48 h after saffron flower harvesting were analyzed for total phenols content (TP) and antioxidant activity (FRAP and DPPH·) to evaluate the effect of post-harvest time on the quality of the floral by-products. Juices obtained by cold-pressing the by-products at 48 h after harvesting exhibited an increase of 61, 56, and 76% for TP, FRAP and DPPH·, respectively, in regard to the values determined for the juices obtained at 24 h after harvesting (4616.1 ± 133.9 mg GAE/L, 35.5 ± 1.4 mmol Fe^{2+}/L, and 7.0 ± 0.3 mmol TEAC/L, respectively for TP, FRAP and DPPH·). These values are about 2 to 3 times higher than the values reported for red wines or berries extracts (Tuberoso et al., 2013).

The main wastes generated by the coffee roasting industry are the roasted silver skin (SS), a tegument of coffee beans usually collected at the roaster exhaust, and the spent coffee grounds (SCG), the exhausted grounds reminiscent of the coffee extraction in the soluble coffee industry. Regazzoni et al. (2016) investigated the feasibility of using SS and SCG as a source of chlorogenic acids, a group of cinnamic acid derivatives with known antioxidant activities. The antioxidant activity of the extracts from SS and SCG was evaluated by means of the DPPH• and ORAC assays. The powdered materials were extracted with a solution of ethanol/water (70:30, v/v) for 18 hours at room temperature, under mechanical stirring, subsequently filtered under vacuum and the filtrate was concentrated with a rotary evaporator at 40°C under reduced pressure. The extraction yields were 13-15% w/w and 4.91% w/w for SS and SCG, respectively, and the total polyphenol contents were higher for SCG than for SS. However, considering chlorogenic acids in free form, SS was a richer source than SCG, with the total amount of chlorogenic acids in the corresponding extracts being reduced by only 38% when considering the starting material (green coffee beans). Although, the chlorogenic acids content of SCG was about 10% of that of the starting material, SCG still maintained a significant antioxidant activity since it contained a significant amount of bound phenolics.

2.2. Recovery of Antioxidants for Application as Food Preservatives

Breweries waste stream contain high amounts of polyphenols, resulting from their removal by a polyvinylpolypyrrolidone resin (PVPP) employed to stabilize beer and extend its shelf life, and these polyphenols can be recovered from the PVPP by an alkaline treatment (Mitchell et al., 2005). The high antioxidant activity of the alkaline extract has been demonstrated by Barbosa-Pereira et al. (2013) to be related to the high concentrations of phenolic compounds, such catechins, and gallic, caffeic, p-coumaric and ferulic acids, which act as free radical acceptors and chain breakers. Based on these properties, Barbosa-Pereira et al. (2014) performed a comparative evaluation of the antioxidant effects of PVPP-WS and rosemary extracts and of two synthetic antioxidants (BHT and PG) on the oxidative stability of beef during cold storage. The natural extracts were used to coat active films with antioxidant properties. The natural extract obtained from the brewery waste stream (PVPP-WS extract) exhibited an antioxidant activity seven times higher than that of BHT and six times higher than that of the rosemary extract. The antioxidant effect of the different antioxidants was determined in beef samples by their direct application to the surface of the beef and by their incorporation into a packaging film. When added directly to the surface of the beef samples, both the PVPP-WS and the rosemary extracts proved to be effective antioxidants by significantly reducing lipid oxidation during beef storage, and thus the need for direct contact of the antioxidant extracts with foodstuff was deemed mandatory. The produced active films coated with natural extracts enhanced oxidative stability of beef relative to that obtained with the film control during cold storage, with the extract obtained from the brewery residual stream (PVPP-WS) presenting the best results, in which lipid oxidation was reduced by up to 90%.

High amounts of wastes are annually produced by the crustacean processing industry, and these wastes are mainly comprised of cephalothorax and cuticles from peeling, spoiled muscle (as a result of extensive melanosis or long frozen storage), and cooking juices. The recovery of biopolymers to develop edible films or coatings for food packaging applications is a promising option for a profitable exploitation this waste. Although chitosan has been the major biopolymer extracted from crustacean waste for the production of edible films, Gómez-Estaca et al. (2014) have recently demonstrated the feasibility of developing edible films from shrimp muscle proteins. In a further study, considering that cephalothorax and cuticles from shrimp waste contain other bioactive compounds of interest, such as astaxanthin, and that the tomato industry generates considerable amounts of waste rich in carotenoids (e.g., lycopene), Gómez-Estaca et al. (2015) developed and characterized shrimp muscle protein antioxidant edible films incorporating carotenoids from a shrimp-waste extract rich in astaxanthin and from a tomato peel waste extract rich in lycopene. Significant differences in antioxidant activity were observed between the shrimp muscle proteins and tomato peels lipid extracts and commercial β-carotene, with the food wastes lipid extracts presenting higher activities. These differences were attributed to the presence of other antioxidants in the lipid extracts, for example, α-tocopherol and hydrophobic amino acids. The carotenoid-incorporated films were tested for carotenoid release in 95% ethanol and the film that released the highest amount of carotenoids at the beginning of storage was the one produced from shrimp muscle protein extract. Regarding film antioxidant activity, the shrimp-extract film exerted the highest activity followed by the tomato-extract and the commercial β-carotene extract films. Carotenoid degradation was used as a parameter to analyze the stability of the produced films during storage and astaxanthin showed the slowest degradation during storage among the studied carotenoids. A gradual reduction of released carotenoids was observed as storage continued and the this reduction was consistent with the degradation rate observed for each carotenoid, with the reduction being faster and more intensive for β-carotene than for astaxanthin. However, the results for carotenoid release from the tomato-extract film did not agree with the carotenoid degradation one, with the difference being attributed to a possible interaction of lycopene with the film matrix, thus hindering its release. Nevertheless, degradation of lycopene was also observed. The higher stability of astaxanthin was attributed to the higher antioxidant activity of the shrimp extract and to its presence in the form of mono- and diesters rather than free astaxanthin. Both tomato-extract lycopene and commercial β-carotene were present in the free form, thus making them more prone to degradation and to interact with proteins in the film. The degradation of carotenoids had a negligible effect on the films' antioxidant activity and thus it remained stable throughout storage. The stability of the films' antioxidant activity during storage was then ascribed to the presence of other antioxidant compounds in both shrimp and tomato extracts and to the presence of degradation products with antioxidant activity, as deduced from the stable behavior of the commercial β-carotene film. Thus, the development of edible antioxidant films incorporating lipid extracts rich in carotenoids was deemed a feasible way to valorize shrimp and tomato waste products.

The trends in the use of natural antioxidants in active food packaging has been recently reviewed by Sanches-Silva et al. (2014). In their review, a few works were discussed in which the active antioxidant components added to the food package were extracted from food waste and they are briefly outlined next. Song et al. (2012) developed a barley bran protein–gelatin composite film containing an antioxidant and antimicrobial grapefruit seed extract, and successfully applied it to salmon packaging. In a series of works, Pereira de Abreu et al. (2010, 2011a,b,c, 2012) evaluated and successfully confirmed the effectiveness of an active packaging film containing antioxidants extracted from barley husks, an agricultural waste, to protect salmon, halibut, hake, blue shark and cod against oxidative lipid degradation during storage. Sanches-Silva et al. (2013) reported the successful incorporation of astaxanthin, obtained from shrimp by-products, in low-density polyethylene matrix. LDPE films, with different amounts of shrimp by-product astaxanthin, were prepared by extrusion and were further evaluated for astaxanthin migration from the plastic matrix into fatty food simulants. Astaxanthin migration from LDPE films into the tested fatty food simulants was not detected by ultra-performance liquid chromatography.

Jully et al. (2016) evaluated the possibility of adding spent coffee grounds, together with whole ground and lyophilized coffee brew, to precooked pork in order to extend the product shelf life, based on the residue antioxidant activity. Light and dark roasted spent coffee grounds were added and Malondialdehyde (MDA), a secondary product of lipid oxidation, and hexanal were monitored for three months. Spent coffee ground, whole ground, and lyophilized brew lowered MDA and hexanal in frozen cooked pork patties to the same extent as rosemary oleoresin, demonstrating great potential as an alternative antioxidant for protection against lipid oxidation. Addition of coffee, in all forms, had no impact on protein oxidation of the pork product, as measured by free thiol content or metmyoglobin formation. All forms of both coffee roast degrees added to the cooked pork patties were able to extend the shelf life of the frozen product.

2.3. Recovery of Antioxidants for Applications Not Related to Food

To consider the recovery of polyphenols for exclusive applications in food would be a shortsighted interpretation of their potential profitable applications as antioxidant compounds. Hence, organic molecules extracted from food waste have been evaluated as alternative steel corrosion inhibitors due to their biodegradability and easy availability. Various food waste extracts have been successfully tested as corrosion inhibitors of steel in acidic media, such as fruit and garlic peel extracts (Rocha et al., 2010; Pereira et al., 2012; Behpoura et al., 2012; Ji et al., 2015), with the corrosion inhibition ability of plant extracts being generally attributed to the presence of antioxidant polyphenolic compounds constituents like alkaloids, flavonoids, or condensed tannins (Behpoura et al., 2012). Orange peel, a waste from the juice extraction process, constitutes almost 50% of the total fruit mass and this waste constitutes a valuable source of phenolic compounds (Rezzadori et al., 2012) exhibiting significant antioxidant activities. In view of this, the corrosion inhibition efficiency of maltaise orange peel extract on carbon steel in hydrochloric acid was investigated by M'hiri et al. (2016). The main phenolic compounds in the orange peel extracts were identified by HPLC as neohesperidin, hesperidin, nobiletin, naringin, narirutin and ascorbic acid. Antioxidant activity was evaluated for the orange peel extract and for the main individual molecules identified in the extract. The orange peel extract presented higher antioxidant activity than expected from the addition of individual contributions from the major phenolic compounds. This difference in antioxidant activity was attributed to the presence of other natural antioxidants, such as carotenoids, and to synergistic effects between different antioxidant molecules, particularly between flavonoids and ascorbic acid. The corrosion inhibition efficiency of the orange peel extract and of selected phenolic compounds (neohesperidin, naringin) as well as the one of ascorbic acid was characterized by using electrochemical methods and was compared to their antioxidant activity as evaluated by ABTS assay. Regardless of concentration, a significant corrosion inhibition was observed by the orange peel extract, whereas selected antioxidant compounds exhibited only a slight inhibition effect. The electrochemical impedance spectroscopy results together with scanning electron microscopy observations after immersion revealed that the inhibiting efficiency of the orange peel extract was not only due to the antioxidant activity of its compounds but also to the precipitation of a surface film. Inhibition of cathodic reaction was partly explained by the antioxidant activity exerted by the presence of ene-diol or catechol functions or other derived functions. Anodic inhibition was attributed to the same kinds of functions because of their chelating properties regarding metallic cations. In presence of extract, the corrosion rate was actually reduced by a progressive precipitation of a covering film, comprised of compounds extracted from the orange peel (e.g., polysaccharides), which could cover the steel surface and trap the antioxidant molecule chelating the Fe^{2+} cations.

CONCLUSION

Agricultural and food wastes are generated annually in significant amounts and constitute rich and cheap sources of bioactive compounds and functional food ingredients. Polyphenols, although not the only class of compounds with desirable antioxidant activities, are probably the most researched compounds for this type of property and have been demonstrated to be quite abundant in the plant kingdom and consequently in agricultural and food wastes. Thus, under the claim of health benefits, the majority of studies on the recovery of antioxidants from agricultural and food wastes have been directed towards their application as functional food ingredients, establishing the profile of phenolics and correlated antioxidant activity of a given waste. Another interesting possibility is the application of the recovered phenolics in food preservation, either as additives applied directly to the food product or as a component of bioactive food packages. Regardless of the application mode, the efficacy of this class of compounds in protecting food products against oxidative damages has been many times proven. Synergistic effects of distinct phenolic compounds, regarding antioxidant activity, have been several times demonstrated more effective in the prevention of lipid oxidation than those of the individual phenolics comprising the food waste extracts.

Recently, a few works published in the literature have claimed the health effects of polyphenols in diets not to be well established and there are even statements claiming that certain amounts of polyphenols may be prejudicial. Although there exists some controversy, the majority of work body published in the literature present positive correlations between consumption of dietary polyphenols and health benefits and, thus, the recovery of polyphenols from agricultural and food waste still constitutes a relevant option for the profitable exploitation of these materials. Review articles in the recovery of bioactive compounds from food wastes claim that existing technologies are suitable for this purpose. However, the recovery of purified polyphenols from food wastes has been proven difficult, since they are present in free and bound forms in the biological matrix, and the development of efficient techniques for the extraction, separation and purification of polyphenols from wastes constitutes a major issue to be properly addressed.

Antioxidant activity is not a property of exclusive interest to the food industry and recently applications of antioxidant polyphenols recovered from food wastes such as steel corrosion inhibitors have been intensively studied. Other non-food applications of great relevance would be the use of waste-recovered polyphenols as antioxidants for the protection against oxidative degradation of fuels, such as biodiesel, and for the protection of polymeric materials that are used in oxidative environment.

ACKNOWLEDGMENTS

The authors gratefully acknowledge financial support from the following Brazilian Government Agencies: CNPq and FAPEMIG.

References

Allouche, N., Fki I., Sayadi, S. (2004) Towards a high yield recovery of antioxidants and purified hydroxytyrosol from olive mill wastewaters, *Journal of Agricultural and Food Chemistry,* 52, 267–273.

Ambigaipalan, P., Shahidi, F. (2015) Antioxidant Potential of Date (Phoenix dactylifera L.) Seed Protein Hydrolysates and Carnosine in Food and Biological Systems, *Journal of Agricultural and Food Chemistry,* 63, 864–871.

André, C., Castanheira, I., Cruz, J.M., Paseiro, P., Sanches-Silva, A. (2010) Analytical strategies to evaluate antioxidants in food: a review, *Trends in Food Science and Technology,* 21, 229-246.

Angelino, D., Gennari, L., Blasa, M., Selvaggini, R., Urbani, S., Esposto, S., Servili, M., Ninfali, P. (2011) Chemical and cellular antioxidant activity of phytochemicals purified from olive mill waste waters, Journal of Agricultural and Food Chemistry, 59, 2011–2018.

Barbosa-Pereira, L., Angulo, I., Paseiro-Losada, P., Cruz, J.M. (2013) Phenolic profile and antioxidant properties of a crude extract obtained from a brewery waste stream, *Food Research International,* 51, 663–669.

Barbosa-Pereira, L., Aurrekoetxea, G.P., Angulo, I., Paseiro-Losada, P., Cruz, J.M. (2014) Development of new active packaging films coated with natural phenolic compounds to improve the oxidative stability of beef, *Meat Science,* 97, 249–254.

Behpoura, M., Ghoreishia, S.M., Khayatkashania, M., Soltanib, N. (2012) Green approach to corrosion inhibition of mild steel in two acidic solutions by the extract of Punica granatum peel and main constituents, *Mater. Chem. Phys.,* 131, 621–633.

Bernini, R., Mincione, E., Barontini, M., Crisante, F., (2008) Convenient synthesis of hydroxytyrosol and its lipophilic derivatives from tyrosol or homovanillyl alcohol, *Journal of Agricultural and Food Chemistry,* 56, 8897–8904.

Blancas-Benitez, F.J., Avena-Bustillos, R.J., Montalvo-González, E., Sáyago-Ayerdi, S.G., McHugh, T.H. (2015) Addition of dried 'Ataulfo' mango (Mangifera indica L) by-products as a source of dietary fiber and polyphenols in starch molded mango snacks, *Journal of Food Science and Technology,* 52, 7393–7400.

Blidi, S., Bikaki, M., Grigorakis, S., Loupassaki, S., Makris, D.P. (2015) A comparative evaluation of bio-solvents for the efficient extraction of polyphenolic phytochemicals: Apple waste peels as a case study, *Waste Biomass Valor,* 6, 1125–1133.

Cardinali, A., Cicco, N., Linsalata, V., Minervini, F., Pati, S., Pieralice, M., Tursi, N., Lattanzio, V. (2010) Biological activity of high molecular weight phenolics from olive mill wastewater, *Journal of Agricultural and Food Chemistry,* 58, 8585–8590.

Cardinali, A., Pati, S., Minervini, F., D'Antuono, I., Linsalata, V., Lattanzio, V. (2012) Verbascoside, isoverbascoside, and their derivatives recovered from olive mill wastewater as possible food antioxidants, *Journal of Agricultural and Food Chemistry,* 60, 1822−1829.

El-Abbassi, A., Kiai, H., Hafidi, A. (2012) Phenolic profile and antioxidant activities of olive mill wastewater, *Food Chemistry,* 132, 406–412.

Ena, A., Pintucci, C., Carlozzi, P. (2012) The recovery of polyphenols from olive mill waste using two adsorbing vegetable matrices, *Journal of Biotechnology,* 157, 573–577.

Federici, F., Fava, F., Kalogerakisc, N., Mantzavinosc, D. (2009) Valorization of agro-industrial by-products, effluents and waste: concept, opportunities and the case of olive mill wastewaters, *Journal of Chemical Technology and Biotechnology,* 84, 895–900.

Galanakis, C.M., Tornberg, E., Gekas, V. (2010) Recovery and preservation of phenols from olive waste in ethanolic extracts, *Journal of Chemical Technology and Biotechnology,* 85, 1148–1155.

Galanakis, C.M. (2012) Recovery of high added-value components from food wastes: Conventional, emerging technologies and commercialized applications, *Trends in Food Science and Technology,* 26, 68-87.

Galanakis, C.M. (2013) Emerging technologies for the production of nutraceuticals from agricultural by-products a viewpoint of opportunities and challenges, *Food and Bioproducts Processing,* 91, 575–579.

García-Lomillo, J., González-SanJosé, M.L., Pino-García, R., Rivero-Pérez, M.D., Muñiz-Rodríguez, P., (2014) Antioxidant and antimicrobial properties of wine byproducts and their potential uses in the food industry, *Journal of Agricultural and Food Chemistry,* 62, 12595−12602.

Girotto, F., Alibardi, L., Cossu, R. (2015) Food waste generation and industrial uses: A review, *Waste Management,* 45, 32–41.

Gómez, A., López, R., Esturo, A., Bald, C., Tueros, I., Talens, C., Raynaud, C. (2015) From waste products to raw materials for the development of new foods, *Waste and Resource Management,* 168, Issue WR2, 55–62.

Gómez-Estaca, J., Montero, P. and Gómez-Guillén, M.C. (2014). Shrimp (Litopenaeus vannamei) muscle proteins as source to develop edible films, *Food Hydrocolloids,* 41, 86-94.

Gómez-Estaca, J., Calvo, M.M., Sánchez-Faure, A., Montero, P., Gómez-Guillén, M.C. (2015) Development, properties, and stability of antioxidant shrimp muscle protein films incorporating carotenoid-containing extracts from food by-products, *LWT - Food Science and Technology,* 64, 189-196.

Gortzi, O., Lalas, S., Chatzilazarou, A., Katsoyannos, E., Papaconstandinou, S., Dourtoglou, E., (2008) Recovery of natural antioxidants from olive mill wastewater using Genapol-X080, *Journal of the American Oil Chemistry Society,* 85, 133–140.

Halliwell, B., Gutteridge, J. (2007) Free radicals in biology and medicine. 4th Ed. Oxford: Oxford University Press Inc., 851 p.

Haq, Z., Easterly, J.L. (2006) Agricultural residue availability in the United States, *Applied Biochemistry and Biotechnology,* 129-132, 3-21.

Ji, G., Anjum, S., Sundaram, S., Prakash, R. (2015) Musa paradisica peel extract as green corrosion inhibitor for mild steel in HCl solution, *Corrosion Science,* 90, 107–117.

Jully, K.M.M., Toto, C.S., Were, L. (2016) Antioxidant effect of spent, ground, and lyophilized brew from roasted coffee in frozen cooked pork patties, *LWT - Food Science and Technology,* 66, 244–251.

Katsampa, P., Valsamedou, E., Grigorakis, S., Makris, D.P. (2015) A green ultrasound-assisted extraction process for the recovery of antioxidant polyphenols and pigments from onion solid wastes using Box-Behnken experimental design and kinetics, *Industrial Crops and Products,* 77, 535–543.

Khiari, Z., Makris, D.P., Kefalas, P. (2009) An investigation on the recovery of antioxidant phenolics from onion solid wastes employing water/ethanol-based solvent systems, *Food Bioprocess Technol.,* 2, 337–343.

Kiassos, E., Mylonaki, S., Makris, D.P., Kefalas, P. (2009) Implementation of response surface methodology to optimise extraction of onion (Allium cepa) solid waste phenolics, *Innov. Food Sci. Emerg. Technol.,* 10, 246–252.

Kris-Etherton, P.M., Hecker, K.D., Bonanome, A., Coval, S.M., Binkoski, A.E., Hilpert, K.F., Griel, A.E., Etherton, T.D. (2002) Bioactive compounds in Foods: Their role in the prevention of cardiovascular disease and cancer, *American Journal of Medicine,* 113, 71–88.

Lafka, T.I., Lazou, A.E., Sinanoglou, V.J., Lazos, E.S. (2011) Phenolic and antioxidant potential of olive oil mill wastes, *Food Chemistry,* 125, 92–98.

Lal, R. (2008) Crop residues as soil amendments and feedstock for bioethanol production, *Waste Management,* 28, 747–758.

Laufenberg, G., Kunz, B., Nystroem, M. (2003) Transformation of vegetable waste into value added products: (A) the upgrading concept; (B) practical implementations, *Bioresource Technology,* 87, 167–198.

Lee, S.D., Yoo, G., Chae, H.J., In, M.J., Oh, N.S., Hwang, Y.K., Hwang, W.I., Kim, D.C. (2009) Lipid-soluble extracts as the main source of anticancer activity in ginseng and ginseng marc, *J. Am. Oil Chem. Soc.,* 86, 1065-1071.

Lee, C.Y. (2013) Challenges in providing credible scientific evidence of health benefits of dietary polyphenols, *Journal of Functional Foods,* 5, 524-526.

Lee, S.H., Oh, M., Park, J., Jang, S.Y., Cheong, S.H., Lee, H., Moon, S.H. (2015) Antioxidant and anti-inflammatory activities of the ethanolic extract of fermented red ginseng marc, *Food Science and Biotechnology,* 24, 651-657.

Lora, E.S., Andrade, R.V. (2009) Biomass as energy source in Brazil, *Renewable and Sustainable Energy Reviews,* 13, 777-788.

Makris, D.P. (2010) Optimisation of anthocyanin recovery from onion (Allium cepa) solid wastes using response surface methodology, *Journal of Food Technology,* 8, 183–186.

M'hiri, N., Veys-Renaux, D., Rocca, E., Ioannou, I., Boudhrioua, N.M., Ghoul, M. (2016) Corrosion inhibition of carbon steel in acidic medium by orange peel extract and its main antioxidant compounds, *Corrosion Science,* 102, 55–62.

Mishra, K., Ojha, H., Chaudhury, N.K. (2012) Estimation of antiradical properties of antioxidants using DPPH· assay: A critical review and results, *Food Chemistry,* 130, 1036–1043.

Mitchell, A.E., Hong, Y.J., May, J.C., Wright, C.A., Bamforth, C.W. (2005) A comparison of polyvinylpolypyrrolidone (PVPP), silica xerogel and a polyvinylpyrrolidone (PVP)–silica co-product for their ability to remove polyphenols from beer, *Journal of the Institute of Brewing,* 111, 20–25.

Mustafa, A.M., Maggi, F., Öztürk, N., Öztürk, Y., Sagratini, G., Torregiani, E., Vittori, S., Caprioli, G. (2016) Chemical and biological analysis of the by-product obtained by processing Gentiana lutea L. and other herbs during production of bitter liqueurs, *Industrial Crops and Products,* 80, 131–140.

Pereira de Abreu, D.A., Paseiro Losada, P., Maroto, J., Cruz, J.M. (2010) Evaluation of the effectiveness of a new active packaging film containing natural antioxidants (from barley husks) that retard lipid damage in frozen Atlantic salmon (Salmo salar L.), *Food Research International,* 43, 1277–1282.

Pereira de Abreu, D.A., Paseiro Losada, P., Maroto, J., Cruz, J.M. (2011a) Lipid damage during frozen storage of Atlantic halibut (Hippoglossus hippoglossus) in active packaging film containing antioxidants, *Food Chemistry,* 126, 315–320.

Pereira de Abreu, D.A., Paseiro Losada, P., Maroto, J., Cruz, J.M. (2011b) Lipid damage inhibition in hake by active packaging film with natural antioxidants, *Packaging Technology and Science,* 24, 353–360.

Pereira de Abreu, D.A., Paseiro Losada, P., Maroto, J., Cruz, J.M. (2011c) Natural antioxidant active packaging film and its effect on lipid damage in frozen blue shark (Prionace glauca), *Innovative Food Science and Emerging Technology,* 12, 50–55.

Pereira de Abreu, D.A., Maroto, J., Rodríguez, K.V., Cruz, J.M. (2012) Antioxidants from barley husks impregnated in films of low density polyethylene and their effect over lipid deterioration of frozen cod (Gadus morhua), *Journal of The Science of Food and Agriculture,* 92, 427–432.

Pereira, S.S.A.A., Pegas, M.M., Fernandez, T.L., Magalhaes, M., Schöntag, T.G., Lago, D.C., Senna, L.F., D'Elia, E. (2012) Inhibitory action of aqueous garlic peel extract on the corrosion of carbon steel in HCl solution, *Corrosion Science,* 65, 360–366.

Purohit, P., Tripathi, A.K., Kandpal, T.C. (2006) Energetics of coal substitution by briquettes of agricultural residues, *Energy,* 31, 1321-1331.

Regazzoni, L., Saligari, F., Marinello, C., Rossoni, G., Aldini, G., Carini, M., Orioli, M. (2016) Coffee silver skin as a source of polyphenols: High resolution mass spectrometric profiling of components and antioxidant activity, *Journal of Functional Foods,* 20, 472–485.

Rezzadori, K., Benedetti, S., Amante, E.R. (2012) Proposals for the residues recovery: Orange waste as raw material for new products, *Food Bioprod. Process.,* 90, 606–614.

Rocha, J.C., Gomes, J.C.C.P., D'Elia, E. (2010) Corrosion inhibition of carbon steel in hydrochloric acid solution by fruit peel aqueous extracts, *Corrosion Science,* 52, 2341–2348.

Sanches-Silva, A., Costa, D., Albuquerque, T.G., Buonocore, G.G., Ramose, F., Castilho, M.C., Machado, A.V., Costa, H.S. (2014) Trends in the use of natural antioxidants in active food packaging: a review, *Food Additives and Contaminants: Part A,* 31, 374–395.

Santana-Méridas, O., González-Coloma, A., Sánchez-Vioque, R. (2012) Agricultural residues as a source of bioactive natural products, *Phytochemical Reviews,* 11, 447–466.

Schieber, A., Stintzing, F.C., Carle, R. (2001) By-products of plant food processing as a source of functional compounds — recent developments, *Trends in Food Science and Technology,* 12, 401–413.

Shahidi, F., Ambigaipalan, P. (2015) Phenolics and polyphenolics in foods, beverages and spices: Antioxidant activity and health effects – A review, *Journal of Functional Foods,* 18, 820–897.

Silva, A.C., Jorge, N. (2014) Bioactive compounds of the lipid fractions of agro-industrial waste, *Food Research International,* 66, 493–500.

Soquetta, M.B., Stefanello, F.S., Huerta, K.M., Monteiro, S.S., Rosa, C.S., Terra, N.N. (2016) Characterization of physiochemical and microbiological properties, and bioactive compounds, of flour made from the skin and bagasse of kiwi fruit (Actinidia deliciosa), *Food Chemistry,* 199, 471–478.

Song, H.Y., Shin, Y.J., Song, K.B. (2012) Preparation of a barley bran protein–gelatin composite film containing grapefruit seed extract and its application in salmon packaging, *Journal of Food Engineering,* 113, 541–547.

Tana, H., Li, H., Song, H., Xua, W., Guana, C., Ran, L. (2012) A novel way of separation and preparation noncaffeine tea polyphenols from green tea waste, *Advanced Materials Research,* 550-553, 1875-1880.

Tuberoso, C.I.G., Boban, M., Bifulco, E., Budimir, D., Pirisi, F.M. (2013) Antioxidant capacity and vasodilatory properties of Mediterranean food: The case of Cannonau wine, myrtle berries liqueur and strawberry-tree honey, *Food Chemistry,* 140, 686–691.

Tuberoso, C.I.G., Rosa, A., Montoro, P., Fenu, M.A., Pizza, C. (2016) Antioxidant activity, cytotoxic activity and metabolic profiling of juices obtained from saffron (Crocus sativus L.) floral by-products, *Food Chemistry,* 199, 18–27.

Uribe, E., Pasten, A., Lemus-Mondaca, R., Vega-Gálvez, A., Quispe-Fuentes, I., Ortiz, J., Di Scala, K. (2015) Comparison of chemical composition, bioactive compounds and antioxidant activity of three olive-waste cakes, *Journal of Food Biochemistry,* 39, 189–198.

Visioli, F., Romani, A., Mulinacci, N., Zarini, S., Conte, D., Vincieri, F.F., Galli, C. (1999) Antioxidant and other biological activities of olive mill waste waters, *Journal of Agricultural and Food Chemistry,* 47, 3397-3401.

Young, I.S., Woodside, J.V. (2001) Antioxidants in health and disease, *Journal of Clinical Pathology,* 54, 176–186.

Yu, J., Ahmedna, M. (2013) Functional components of grape pomace: their composition, biological properties and potential applications, *International Journal of Food Science and Technology,* 48, 221–237.

In: Food Waste
Editor: Garrett Leonard Riley

ISBN: 978-1-63485-025-4

Chapter 3

POTENTIAL OF SPENT COFFEE GROUNDS AS SOURCES OF DIETARY FIBER WITH ANTIOXIDANT ACTIVITY

Leandro S. Oliveira*[*] *and Adriana S. Franca
Universidade Federal de Minas Gerais
Belo Horizonte, Brasil

ABSTRACT

Coffee is one of the most popular and consumed beverages over the world and thus quite important as a food commodity from an economic point of view. Both the production of coffee beverage or soluble coffee generate significant amounts of solid residues, known as spent coffee grounds (SCG). Such residues do not present any commercial value and are in most part disposed either in landfill sites or by burning as fuel in the boilers of the soluble coffee industry. However, they are highly pollutant due to high contents of organic substances that demand great quantities of oxygen to decompose. This has stimulated efforts to find ways of reducing their environmental impact and/or transforming them into value-added products. Some of the evaluated applications include production of fertilizers, biofuels, and use as substrates for the cultivation of mushrooms or for the production of adsorbents targeting wastewater treatment. Another promising but still relatively unexplored approach is the use of SCG as a raw material for the recovery of functional

[*]Email: leandro@demec.ufmg.br.Tel:+55-31-34093517. Fax:+55-31-34433783.

compounds of potential interest to the food industry. Coffee phenolics have attracted much interest in recent years due to their strong antioxidant and metal-chelating properties, and thus are believed to provide protection against free radical damage and reduce the risk of degenerative diseases associated with oxidative stress. Antioxidants are key ingredients in functional foods, given their well-established health benefits. These compounds are usually available in food wastes, mainly those derived from fruit husks and seeds that have been also evaluated as alternative sources of dietary fibers (DF). The well-established health benefits associated with DF have been also associated to the presence of phenolics. SCG can thus be viewed as a potential source of DF with antioxidant capacity. In view of the aforementioned, the objective of the present study is to present an overview of the research works that have been developed in the last decade on the alternative uses of spent coffee grounds. A critical overview on the studies that deal specifically with food applications is presented, mainly focusing on their potential as sources of fibers and associated phenolics and respective antioxidant activity.

1. INTRODUCTION

Spent coffee grounds correspond to roasted and ground coffee depleted of water solubles. They represent the solid residues that are generated after the coffee beverage is prepared and are available at households, commercial establishments or from the production of soluble coffee. SCG are disposed of either in landfill sites or by burning as fuel in the boilers of the soluble coffee industry. However, they present several environmental problems in terms of appropriate disposal. These solid residues are highly pollutant due to high contents of organic substances that demand great quantities of oxygen to decompose (Franca and Oliveira, 2009). Also, spontaneous combustion can occur and has been reported in some storage sites (Silva et al., 1998). Besides the environmental implications, SCG presents an additional disposal problem, given that they can be used for adulteration of roasted and ground coffee, being very difficult to detect as an adulterant (Reis et al., 2013ab). Thus, the soluble coffee industry has been quite careful with its disposal, and most of the time this residue is simply used as a boiler fuel by the same industry.

As it is common for agricultural and food wastes in general, the first attempts towards finding alternative uses for SCG have been concentrated on applications as fuel (Adams and Dougan, 1987), fertilizers (Tango, 1971), and supplements for animal feed (Sikka et al., 1985; Givens and Barber, 1986;

Sikka and Chawla, 1986). Among these applications, recent studies have confirmed its potential use as a solid fuel (Xu et al., 2006; Horio et al., 2009; Limousy et al., 2013; Gómez-de la Cruz et al., 2015; Limousy et al., 2015) and in composting (Hachicha et al., 2012; Cruz et al., 2014ab, 2015). Other recent applications include the production of adsorbents (Franca et al., 2009; Plaza et al., 2012; González et al., 2013) and the production of liquid biofuels (Kwon et al., 2013; Rocha et al., 2014).

Coffee is one of the most widely consumed pharmacologically active beverages and moderate daily consumption of coffee has been associated to several positive health effects (Bae et al., 2014). Some studies have demonstrated that coffee consumption presents inverse correlation with several diseases including Alzheimer's, cancer, Parkinson's, diabetes and depression, among others (Higdon and Frei, 2006; Nkondjock, 2009; Butt and Sultan, 2011; Lucas et al., 2011; Hall et al., 2015). Coffee also shows protective effects on various systems including the skeletal system, the reproductive system, the nervous system and the cardiovascular system (George et al., 2008). Such health-related benefits are attributed to coffee's rich phytochemistry, including substances such as caffeine (the most widely used psychoactive substance) and other biologically active compounds predominantly belonging to the polyphenol and alkaloid classes (Hall et al., 2015). These substances are present not only in coffee beans or the associated beverage, but also in solid wastes generated during coffee processing, particularly spent coffee grounds (SCG).

In view of the aforementioned, recent studies have focused on the characterization of spent coffee grounds as sources of bioactive compounds. In the present chapter, we present an overview of such studies, in order to better establish the functional potential of spent coffee grounds.

2. Chemical Composition

The average proximate composition of spent coffee grounds is expected to be similar to that of roasted coffee beans, given that SCG correspond to the solid residues remaining after water extraction of soluble components. A comparative evaluation of the chemical composition of both SCG and roasted coffee is presented in Table 1. No significant differences can be observed in terms of protein levels. However, reported levels are based on total nitrogen compound determination and thus may be overestimated due to the presence of other nitrogen-containing substances such as caffeine, trigonelline, free amines and amino acids (Delgado et al., 2008). This class of compounds is relatively stable between species, and also during roasting, although several nitrogen based substances will be extracted during beverage preparation.

Table 1. Chemical composition roasted coffee and spent coffee grounds (g/100 g dry basis)[a]

	Roasted Arabica	Roasted Robusta	Spent coffee grounds
Protein	12-15	13-15	10–17
Lipids	15-20	11-16	22–27
Minerals	4-5	5	0.1-1
Carbohydrate	40-79[b]	64-71[b]	45-89[c]
Caffeine	~1	~2	0.07-0.4

[a] Compilation of data presented by Franca and Oliveira, 2008; 2009; Delgado et al., 2008; Ballesteros et al., 2014 [b] Determined by difference [c] Cellulose + lignin.

Lipid contents in SCG are high in comparison to green and roasted coffee. Given that they are not expected to be extracted during beverage preparation, the higher values are attributed to the fact that the contents are expressed on dry basis of spent coffee grounds and such values are smaller than dry basis of roasted coffee, given the loss of solubles during extraction. This effect was not noticed in protein levels probably due to extraction of nitrogen-based substances like caffeine during beverage preparation. Lower mineral levels in SCG are expected in comparison to roasted coffee, given that most minerals are easily extracted with hot water.

Coffee beans are a rich source of polysaccharides, mainly consisting of mannans or galactomannans, type II arabinogalactans, and celulose (Campos-Veja et al., 2015). Roasting increases both bean arabinogalactan and mannan solubility by loosening the cell-wall structure as the beans swell and by polysaccharide depolymerization, so they can be extracted during beverage preparation. However, most of these polysaccharides remain as insoluble material bound to the SCG matrix (Mussatto et al., 2011a; Simões et al., 2013), thus SCG contain high amounts of sugars polymerized into cellulose and hemicellulose structures. The average sugar composition is reported to be 37-46% mannose, 27-32% galactose, 20-24% glucose, and 7% arabinose (Mussatto et al., 2011a; Ballesteros et al., 2014).

Caffeine concentration in SCG is much lower in comparison to roasted coffee, due to extraction during beverage preparation. Values vary considerably, ranging from 0.7 to over 40 μg/mg. Apart from the differences in caffeine levels among species (arabica contains approximately half of the caffeine in robusta), variations are due mainly to the type of solvent being used, with higher values associated to the use of dichloromethane, water or CO_2 (solvents commonly employed for decaffeination) and lower values if hexane is employed (Ramalakshmi et al., 2009; Andrade et al., 2012; Campos-Vega et al., 2015).

As previously mentioned, SCG are very rich in polysaccharides, which are polymers that can be used as dietary fibre and present immunostimulatory activity (Simões et al., 2013). SCG are primarily composed of neutral detergent fiber (45.2%) occurring as hemicellulose, cellulose, and lignin associated compounds, and acid detergent fiber (29.8%), consisting of cellulose and lignin (Vardon et al., 2013). Reported values for total dietary fiber content range from 45 to 51%, comprised of 35 to 48% insoluble and 2 to 8% soluble (Murthy and Naidu, 2012; Vilela, 2015). Total dietary fiber (TDF) contents are slightly higher than commonly employed fiber sources such as rice and wheat bran (~27-45 g/100 g) (Elleuch et al., 2011) and in the same range of fiber rich powders based on banana flour (~42 g/100 g) (Rodríguez-Ambriz et al., 2008), chia seeds (~41 g/100 g) (Reyes-Caudilo et al., 2008), carrot peels (~45 g/100 g) (Chantaro et al., 2008), and mango peels (~51 g/100 g) (Ajila et al., 2010). TDF values are slightly low in comparison to other coffee processing byproducts such as coffee silverskin (~62 g/100 g) (Borreli et al., 2004) and coffee husks and pulp (~66 g/100 g) (Melo, 2013). However, TDF values in SCG can increase up to 70g/100 g after treatment with hydrogen peroxide (Vilela, 2015).

The soluble (SDF) and insoluble (IDF) nature of dietary fibers will affect not only their technological functionality, but also the physiological effects. IDF are usually associated to porosity, low density and ability to increase faecal bulk and decrease intestinal transit (Elleuch et al., 2011). IDF corresponds to approximately 80 to 90% of the fibers present in SCG. IDF values are in the same range of fruit and vegetable peels such as mango (~39 g/100 g), lemon (~42 g/100 g), orange (~52 g/100 g) and carrots (~34-51 g/100 g), similar to pomegranate bagasse (~29-30 g/100 g), and significantly high in comparison to fruits including apple, orange, dates and tomatoes (~5-12 g/100 g) (Hassan et al., 2011; Elleuch et al., 2011; Viuda-Martos et al., 2012). SDF are also of interest from both technological and functional points of view, in association to their capacity to increase viscosity and to reduce the glycemic response and plasma cholesterol as well as prebiotic action (Elleuch et al., 2011). SDF values in spent coffee grounds are low, in a similar range reported for some fruits such as apple, peach, orange and tomato (~6-10 g/100 g), but quite low in comparison to other agricultural residues such as carrot peel and pomegranate bagasse (~9.8-19.9 g/100 g) (Chantaro et al., 2008; Elleuch et al., 2011;Viuda-Martos et al., 2012). Nonetheless, the coffee fibers obtained from SCG were shown to exhibit antioxidant properties in levels similar to well-known food antioxidant such as red wine products. Thus, dietary fibers from SCG can be categorized as antioxidant dietary fiber, useful as potential dietary supplement (Murthy and Naidu, 2012).

Phenolic compounds are the major substances that are associated to the antioxidant potential of food sources. Phenolic compounds are mainly found in coffee beans as chlorogenic acids (CGA). These correspond to a class of water-soluble esters formed between quinic acid and one or two moieties of caffeic acid, a trans-cinnamic acid. The most important CGA found in coffee include caffeoylquinic, dicaffeoylquinic, feruloylquinic, and *p*-coumaroylquinic acids. Many studies have associated the antioxidant activity and nutraceutical potential of green coffees to CGA (Suzuki et al., 2002; Gawlik-Dziki et al., 2014; Mikami and Yamazawa, 2015). However, CGA levels have been shown to decrease significantly during roasting (up to 75% reduction in concentration) (Wei and Tanokura, 2015). Comparative evaluations on the antioxidant potential of aqueous solutions obtained from green and roasted coffees showed that both green and roasted coffees presented similar antioxidant activities, with a slight decrease for light roasts followed by an increase in darker roasts. This behavior is attributed to the loss of polyphenolic compounds (CGA) during light roasting and to the successive formation of other antioxidant compounds such as Maillard reaction products or pyrolysis products in dark roasts. Therefore, the antioxidant potential of spent coffee grounds is probably due to the combined effect of the remaining CGA as well as Maillard reaction products that were not extracted during beverage preparation. An evaluation of studies that deal with the chemical composition of SCG with respect with substances associated with its antioxidant potential is presented as follows.

3. Recovery of Phenolics and Antioxidant Activity

The recovery of phenolic compounds from SCG as well as their antioxidant activity have been recently investigated (Campos-Vega et al., 2015). Mussatto and co-workers (2011b) evaluated the use of water/methanol mixtures for the recovery of phenolics from spent coffee grounds. Among the evaluated conditions, 90 min extraction employing 60% methanol in a solvent/solid ratio of 40 mL/g of solid was found to be the most efficient for obtaining an extract with high content of phenolic compounds (16 mg gallic acid equivalents/g SCG) and high antioxidant activity (FRAP of 0.10 mM Fe(II)/g). However, the toxic nature of the employed solvent could be a concern in terms of further applications of the recovered phenolics.

Bravo and collaborators (2012) compared SCG obtained from different types of coffeemakers with respect to the concentration of several bioactive substances (3-, 4-, and 5-monocaffeoylquinic and 3,4-, 3,5-, and 4,5-dicaffeoylquinic acids, caffeine, and browned compounds, including melanoidins), and to total phenolics and antioxidant capacity. Evaluated coffeemakers/extraction procedures were filter, espresso, plunger, and mocha. All spent coffee grounds, except for the ones obtained from the mocha coffeemaker, presented significant amounts of total caffeoylquinic acids, ranging from 11.05 (espresso) to 13.24 mg/g (filter) for arabica and from 6.22 (filter) to 7.49 mg/g (espresso) for robusta. The amounts of total caffeoylquinic acids were 4 to 7-fold higher in comparison to the respective coffee brews, indicating that although these compounds will be transferred to the beverage, the majority will remain on the solid matrix and thus contribute to the antioxidant activity of the spent coffee grounds. Caffeine levels ranged from 3.59 to 8.09 mg/g. ABTS-based antioxidant capacity ranged from 8 (mocha) to 215 μmol Trolox/g (filter) for arabica and from 22 (mocha) to 177 μmol Trolox/g (espresso) for robusta. The antioxidant capacities of the aqueous extracts based on SCG represented 46.0–102.3% (filter), 59.2–85.6% (espresso), and <42% (plunger) of the antioxidant capacities of the respective coffee brews. The general conclusion was that SCG obtained from the most commonly employed types of coffeemakers, namely filter and espresso, could be considered as a potential source of hydrophilic bioactive compounds. In a subsequent study Bravo et al. (2013) further evaluated the effects of the extraction procedure on the antioxidant capacity of spent coffee grounds. They evaluated three different extraction methods: Soxhlet apparatus, solid-liquid extraction and conventional extraction using a coffee maker, and three different extractants: water, ethanol, and methanol, pure and mixed. The evaluated parameters were total phenolics (based on Folin-Ciocateau method), antioxidant potential, (based on ABTS and DPPH methods) and browned compounds (based on absorption at 420 nm as a measure of caramelization and Maillard reactions). A comparison of extraction procedures showed that solid-liquid extraction provided extracts with higher amounts of total phenolics and browned compounds, whereas extraction in a coffee maker provided extracts with higher antioxidant capacity. The coffee maker extraction was selected for further evaluation on the basis of antioxidant capacity results. Only one extraction step was deemed sufficient (approx. 80% extraction efficiency) and water was found to be the best solvent. It was also observed that deffating procedures would also improve extraction.

Other studies have focused on optimizing extraction of phenolics from spent coffee grounds. Zuorro and Lavecchia (2012) evaluated the effects of temperature, time, liquid-to-solid ratio and ethanol/water concentration on the recovery of phenolic compounds from SCG. Total phenolics ranged from 17.75 to 21.56 mg GAE/g, with up to 90% recovery. In a later study, Zuorro (2015) employed response surface methodology in order to optimize the recovery of phenolics from spent coffee grounds. All the factors evaluated (temperature, extraction time, liquid to solid ratio and ethanol concentration) had a positive correlation with the extraction of phenolics, although temperature and liquid to solid ratio were the most influential. Maximum extraction yield (97.8%) was obtained under the following optimized conditions: temperature of 47.1°C, 150 min extraction time, liquid/solid ratio of 47.98 mL/g and ethanol concentration of 57.7%. The antioxidant activity of the dry phenolic extract obtained under these conditions was 1583 ± 57 μmol TE/g. Another optimization study was presented by Xu and collaborators (2015). The goal was to maximize both the yield of phenolics and radical scavenging capacity (based on DPPH and ABTS) employing subcritical water as the extractant. Extracts obtained under optimized conditions (179°C, 36 min, and 14.1 g/L) presented the following characteristics: total phenolic content of 86.23 mg GAE/g and scavenging activities of 81.38 (ABTS) and 42.13 (DPPH) mmol TE/100 g. Caffeoylquinic acids were shown to be the major contributors to the antioxidant activities of the produced extracts.

Cruz et al. (2012) evaluated the potential of commercially available SCG from espresso machines as sources of bioactive compounds. The authors reported a high compositional variability, especially with regard to water-soluble components such as caffeine, chlorogenic acids and minerals. There was a strong positive correlation between total soluble solids retained and the levels of caffeine, CGA and minerals, confirming the significance of the type of extraction procedure on the antioxidant potential of SCG. High amounts of CGA (478.9 mg/100 g) and caffeine (452.6 mg/100 g) were reported. The compositional variability of spent coffee grounds was further confirmed in the study of Panusa et al. (2013). Caffeine levels ranged from 599 to 1150 mg/100 g and CGA contents varied from 165 to 609 mg/100 g.

A comprehensive study on the chemical characterization of spent coffee grounds obtained from a soluble coffee industry was presented by Pujol et al. (2013). The evaluated parameters were elemental analysis, mineral composition and ash content, summative composition, acidic functional groups, lipophilic extractives, total polyphenols, and condensed tannins. The spent coffee samples presented high carbon (>58%), low nitrogen (<2%), and low ash (<1%) contents. Lignin and polysaccharides were preset in similar amounts (20 - 26%). The monosaccharide composition was 60% glucose and 40% mannose. The highest values obtained for total polyphenols and tannins were 454 and 293 mg GAE/g, respectively. Assignments of the bands of the obtained FTIR spectra indicated the presence of lipids, polysaccharides and chlorogenic acids.

Pavlović and collaborators (2013) studied the effect of ethanol concentration on the extraction of phenolics from spent coffee grounds. Microwave-assisted extraction was performed and the influence of ethanol concentration and time of microwave radiation was evaluated in terms of extraction yield, total polyphenol content, DPPH radical inhibition activity and FRAP ferric reducing ability. The highest amount of phenolics (398.95 mg GAE/g) was obtained with 20% aqueous ethanol solution under 40 s of microwave radiation. The corresponding IC_{50} value obtained for that extract was 3.75 μg/mL, in the same range of commonly employed commercial antioxidants such as ascorbic acid (~6 μg/mL) and BHT(~3 μg/mL). Another study on microwave-assisted extraction employing low concentration ethanol in aqueous solution was developed by Ranic et al. (2014). They employed response surface methodology in order to monitor the effects of extraction time (ET), liquid-to-solid ratio (LSR), and microwave power (MWP) on total phenolics content and antioxidant activity. Experimental conditions that provided the extract with the highest concentration of phenolics were ET = 40 s, LSR= 6 mL/g and MWP = 240W. FRAP and DPPH-based antioxidant activity showed good correlation with the total polyphenol content (TPC). Experimental conditions that provided the extract with the highest antioxidant activities were ET = 11 s, LSR= 9 mL/g and MWP = 240W.

Acevedo et al. (2013) evaluated different extraction methods (solid–liquid extraction, supercritical extraction or direct saponification) of oil and diterpenes (kahweol and cafestol) from SCG. Both diterpenes have been correlated with health effects, including anti-inflammatory and anticarcinogenic properties (Cavin et al., 2002; Kim et al., 2004; Speer and Kölling-Speer, 2006; Kotowski et al., 2015) The authors also evaluated the phenolic composition and antioxidant capacities of SCG, before and after oil extraction. Among the tested extraction methods, Soxhlet extraction presented the highest oil amounts (26.4%). The free fatty acid profile indicated a high content of polyunsaturated fatty acids (50%), being comprised mainly by linoleic (45%) and palmitic (30%) acids, regardless of the extraction procedure. Results from total phenolics were 255 and 273 mg GAE/g dm for ethanolic extracts obtained from SCG and defatted SCG, respectively. The free radical-scavenging activity was 83 and 201 μmol Trolox/g dm for ethanolic extracts obtained from SCG before and after oil removal, respectively. The highest amounts of diterpenes were extracted by direct saponification (214 mg/100 g SCG for kahweol and 466 mg/100 g SCG for cafestol). Given that polyphenol extraction was not affected by the presence of oil, the authors suggested a sequential extraction process, first extracting the lipid fraction by means of a Soxhlet procedure followed by a polyphenol extraction using an ethanol/water mixture. Therefore, interesting bioactive compounds of interest for the food, cosmetic and pharmaceutical industries could be recovered first from the oil (diterpenes) and then from the solid matrix (phenolics).

Monente et al. (2015a) presented a comparative evaluation of spent coffee grounds and the corresponding coffee brews in terms of composition as well as antimutagenic and antimicrobial activities. Both arabica and robusta coffees were evaluated and also two types of extraction, filter and espresso. Caffeine contents were similar for the SCG and corresponding brews (~ 49 and 80 mg/g for arabica and robusta, respectively) whereas total CGA contents were slightly higher in the spent coffees. CGA values for the spent coffee/corresponding brew were 84/74 and 66/51 mg/g for arabica and robusta, respectively. All samples exhibited strong protection activity against indirect acting mutagen 2-AF (≤92%), whereas the protection against NPD (direct mutagen) was 12–35%. Spent coffee showed antimicrobial activity, mainly against Gram-positive bacteria (*Staphylococcus aureus, Listeria monocytogenes*) and yeast (*Candida albicans*). It was indicated that a high content of melanoidins was responsible for the growth inhibition of Gram-negative bacteria. These results indicate that the composition and the antimutagenic and antimicrobial activity of spent coffee suggest that it can be considered as a potential ingredient for enhancing functional properties and extending the shelflife of food products. The same research group also evaluated the composition of phenolic acids in SCG, both free and bound (Monente et al., 2015b). They performed HPLC analysis of the SCG extracts submitted to three types of treatment: alkaline, acid, saline in order to quantify caffeoylquinic (3-CQA, 4-CQA, 5-CQA), dicaffeoylquinic (3,4-diCQA, 3,5-diCQA, 4,5-diCQA), caffeic, ferulic, *p*-coumaric, sinapic, and 4-

hydroxybenzoic acids. It was observed that actual amount of phenolics present in SCG is over 2-fold higher, in comparison to free compounds. Approximately half of the total phenolics were linked to macromolecules such as melanoidins, mainly by noncovalent interactions. In contrast, coffee brew had only around 20% of attached phenolics Alkaline hydrolysis and saline treatment were found to be adequate for estimation of total bound and ionically bound phenolic acids, respectively. In conclusion, spent coffee extract is a rich source of phenolic acids with a high percentage of compounds linked to macromolecules such as melanoidins or other Maillard reaction products, mainly by noncovalent interactions. Caffeoylquinic acids were found to the most abundant of the phenolics, representing approximately 70% of the total CGAs.

A couple of studies have employed Fourier-transform near infrared spectroscopy (FT-NIR) in order to access the phenolics content and antioxidant potential of spent coffee grounds (Páscoa et al., 2013; Magalhães et al., 2016). Antioxidant values determined in ethanolic extracts ranged from 35 to 296 μmol Trolox/g, total phenolics ranged from 4 to 31 mg GA/g, and total flavonoids ranged from 2 to 20 mg catechin/g (Páscoa et al., 2013). FT-NIR spectra were calibrated against total flavonoid and total phenolic contents and the antioxidant capacity using PLS modeling. The content of three main phenolics (caffeic acid, (+)-catechin and chlorogenic acid) and three methylxanthines (caffeine, theobromine and theophylline) in SCG samples was also evaluated (Magalhães et al., 2016). Results obtained by HPLC coupled with diode-array detection were employed for calibration of the NIR data. The best PLS models were in reference to caffeine content, followed by caffeic acid, (+)-catechin, theophylline and chlorogenic acid, respectively.

A few studies have presented an evaluation of the chemical composition and antioxidant potential of spent coffee grounds in comparison to other coffee processing residues such as coffee husks and pulp and coffee silverskin. The recovery of phenolic compounds from coffee pulp and husk, silver skin, and spent coffee grounds was investigated by Murthy and Naidu (2012). A mixture of isopropanol and water (60:40 v/v) was employed as solvent. Chlorogenic acid content ranged from 2.3 g/100g (spent coffee grounds) to 3.0 g/100g (silverskin). The total polyphenols were 22.19 ± 0.63, 21.71 ± 0.25, 20.12 ± 0.1, and 22 ± 0.2 μg/mL gallic acid equivalents, for coffee pulp, husks, silverskin and spent coffee grounds, respectively. DPPH-based antioxidant activity ranged from 65% to 70% whereas hydroxyl radical scavenging activity (500 ppm) ranged from 59 (coffee pulp) to 85% (spent coffee grounds). Andrade and collaborators (2012) evaluated the chemical composition and the antioxidant activity of spent coffee grounds and coffee husks extracts, obtained by supercritical fluid extraction with CO_2 and with CO_2 and co-solvent (high pressure methods). Low pressure methods (ultrasound and Soxhlet) were also applied to obtain the extracts. The extracts obtained by low pressure extraction with ethanol showed the best results for the global extraction yield and total phenolics (TPC) in comparison to supercritical fluid extraction. The highest TPC values were 151 (coffee husks, Sohxlet) and 588 mg of chlorogenic acid equivalent/g (spent coffee grounds, ultrasound). The highest antioxidant activity was presented by coffee husk extracts obtained by low pressure extraction.

Ballesteros et al. (2014) presented a comparative evaluation of the chemical composition, functional properties, and structural characteristics of coffee silverskin (CS) and spent coffee grounds (SCG). It was concluded that both residues are sugar-rich lignocellulosic materials composed of high levels of insoluble, soluble, and total dietary fibers. The content of total dietary fiber was higher in SCG (60.46% *w/w*) in comparison to CS (54.11% *w/w*). Functional and physiological properties of both residues were similar: water holding capacity (g water/g dry sample) values were 5.73 and 5.11 for SCG and CS, respectively; oil holding capacity (g oil/g dry sample) values were 5.2 and 4.7 for SCG and CS, respectively. Emulsifying activity/emulsion stability values were 54.7/92.4 and 57.5/88.2 for SCG and CS, respectively. The antioxidant potential was similar in terms of DPPH values: 20 and 21 μmol TE/g for SCG and CS, respectively. SCG exhibited twice the antioxidante potential of CS in terms of FRAP results: 0.102 vs. 0.045 mmol Fe(II)/g. Jiménez-Zamora and collaborators (2015) assessed the prebiotic, antimicrobial and antioxidant properties of both spent coffee grounds and coffee silverskin, pure and mixed with melanoidins extracted from spent coffee grounds. All residues were obtained from regular and torrefacto (sugar added) roasted coffees. It was observed that prebiotic activity was important in both types of residue, but that the presence of coffee melanoidins interfered with such beneficial properties. The opposite behavior was reported in terms of antimicrobial activity, that was significantly increased with the addition of melanoidins. All the residues presented antioxidant potential. Addition of melanoidins to the residues as well as addition of sugar during coffee roasting (torrefacto samples) increased the antioxidant and antimicrobial activity.

CONCLUSION

Antioxidants are key ingredients in functional foods, given their well established health benefits. Phenolic compounds are the major substances that are associated to the antioxidant potential of food sources and thus coffee phenolics have attracted much interest in recent years due to their strong antioxidant properties. Phenolic compounds are mainly found in coffee beans as chlorogenic acids and studies have demonstrated that the amount of CGA is much higher in spent coffee grounds in comparison to the corresponding brew, indicating that although these compounds will be transferred to the beverage, the majority will remain on the solid matrix and thus contribute to the antioxidant activity of spent coffee grounds. Therefore, SCG have been the subject of extensive research in order to better evaluate its chemical composition, especially as a source of dietary fiber and phenolics. Total dietary fiber contents have been shown to range from 45 to 51%, values that are higher than commonly employed fiber sources such as rice and wheat bran. SCG present a high compositional variability, especially with regard to water-soluble components such as caffeine, chlorogenic acids and minerals. Reported values for total phenolics and antioxidant potential also vary considerably, with TPC values ranging from 16 to 588 mg GAE/g and antioxidant potential values as low as 8 and as high as 1583 μmol TE/g. Variations can be due to intrinsic compositional variation of the original coffee, but are mainly attributed to differences in extraction procedures (methodologies, solvents, etc.). Studies also indicate that the actual amount of phenolics present in SCG is over 2-fold higher than the reported values, because most of the studies only take into account free compounds. It is expected that approximately half of the total phenolics available in SCG will be linked to macromolecules such as melanoidins. In conclusion, it is well established in the literature that SCG are potential sources of dietary fibers with antioxidant capacity and thus such residues are of great interest for application in the development of food and products.

ACKNOWLEDGMENTS

The authors gratefully acknowledge financial support from the following Brazilian Government Agencies: CNPq and FAPEMIG.

References

Acevedo, F., Rubilar, M., Scheuermann, E., Cancino, B., Uquiche, E., Garcés, M., Inostroza, K. and Shene, C. (2013) Spent coffee grounds as a renewable source of bioactive compounds, *Journal of Biobased Materials and Bioenergy,* 7, 420-428.

Adams, M.R. and Dougan, J. (1987) Waste Products. In: R.J. Clarke and R. Macrae, (Eds.), Coffee Technology, New York: Elsevier Applied Science Publishers Ltd., pp. 257-291.

Ajila, C. M., Aalami, M., Leelavathi, K. and Prasada Rao, U.J.S. (2010) Mango peel powder: A potencial source of antioxidant and dietary fiber in macaroni preparation. Innovative Food Science and Emerging Technologies, 11, 219-224.

Andrade, K.S., Gonçalvez, R.T., Maraschin, M., Ribeiro-do-Valle, R.M., Martínez, J. and Ferreira, S.R.S. (2012) Supercritical fluid extraction from spent coffee grounds and coffee husks: Antioxidant activity and effect of operational variables on extract composition, *Talanta,* 88, 544– 552.

Bae, J.-H., Park, J.-H., Im, S.-S. and Song, D.-K. (2014) Coffee and health. *Integrative Medicine Research*, 3, 189-191.

Ballesteros, L.F. Teixeira, J.A. and Mussatto, S.I. (2014) Chemical, functional, and structural properties of spent coffee grounds and coffee silverskin, *Journal of Food and Bioprocess Technology,* 7, 3493-3503.

Borrelli, C.R., Esposito, F., Napolitano, A., Ritieni, A. and Fogliano, V. (2004) Characterization of a new potential functional ingredient: coffee silverskin. *Journal of Agricultural and Food Chemistry*, 52, 1338-1343.

Bravo, J., Juániz, I., Monente, C., Caemmerer, B., Kroh, L.W., Paz De Peña, M. and Cid, C. (2012) Evaluation of spent coffee obtained from the most common coffeemakers as a source of hydrophilic bioactive compounds, *Journal of Agricultural and Food Chemistry,* 60, 12565–12573.

Bravo, J., Monente, C., Juániz, I. and Paz De Peña, M. and Cid, C. (2013) Influence of extraction process on antioxidant capacity of spent coffee, *Food Research International*, 50, 610-616.

Butt, M.S. and Sultan, M.T. (2011) Coffee and its consumption: Benefits and risks, *Critical Reviews in Food Science and Nutrition,* 51, 363–373

Campos-Vega, R., Loarca-Piña, G., Vergara-Castañeda, H.A. and Oomah, B.D. (2015) Spent coffee grounds: A review on current research and future prospects, *Trends in Food Science and Technology*, 45, 24-36.

Cavin, C., Holzhaeuser, D., Scharf, G., Constable, A., Huber, W.W. and Schilter, B. (2002) Cafestol and kahweol, two coffee specific diterpenes with anticarcinogenic activity, *Food and Chemical Toxicology,* 40, 1155-1163.

Chantaro, P., Devahastin, S. and Chiewchan, N. (2008). Production of antioxidant high dietary fiber powder from carrot peels. *LWT - Food Science and Technology, 4,* 1987-1994.

Cruz, R., Mendes, E., Torrinha, A., Morais, S., Pereira, J.A., Baptista, P. and Casal, S. (2015) Revalorization of spent coffee residues by a direct agronomic approach, *Food Research International,* 73, 190-196.

Cruz, R., Gomes, T., Ferreira, A., Mendes, E., Baptista, P., Cunha, S., Pereira, J.A., Ramalhosa, E. and Casal, S. (2014a) Antioxidant activity and bioactive compounds of lettuce improved by espresso coffee residues, *Food Chemistry,* 145, 95-101.

Cruz, R., S., Mendes, E., Pereira, J.A., Baptista, P. and Casal, S. (2014b) Improvement of vegetables elemental quality by espresso coffee residues, Food Chemistry, 148, 294-299.

Cruz, R., Cardoso, M.M., Fernandes, L., Oliveira, M., Mendes, E., Baptista, P., Morais, S. and Casal, S. (2012) Espresso coffee residues: a valuable source of unextracted compounds, *Journal of Agricultural and Food Chemistry,* 60, 7777–7784.

Delgado, P.A., Vignoli, J.A., Siika-aho, M. and Franco, T.T. (2008) Sediments in coffee extracts: composition and control by enzymatic hydrolysis, *Food Chemistry,* 110, 168–176.

Elleuch, M., Bedigian, D., Roiseux, O., Besbes, S., Blecker, C. and Attia, H. (2011). Dietary fibre and fibre-rich by-products of food processing: Characterisation, technological functionality and commercial applications: A review, *Food Chemistry, 124,* 411-421.

Franca, A.S., Oliveira, L.S., Ferreira, M.E. (2009) Kinetics and equilibrium studies of methylene blue adsorption by spent coffee grounds, *Desalination*, 249, 267-272.

Franca, A. S. and Oliveira, L. S. (2009) Coffee processing solid wastes: current uses and future perspectives. In G.S. Ashworth and P. Azevedo (Eds.), Agricultural Wastes (pp. 155-189), New York: Nova Publishers.

Franca, A. S. and Oliveira, L. S. (2008) Chemistry of Defective Coffee Beans. In E.N. Koeffer (Ed.), Progress in Food Chemistry (pp. 105-138), New York: Nova Publishers.

Gawlik-Dziki, U., Świeca, M., Dziki, D., Kowalska, I., Pecio, Ł., Durak, A. and Sęczyk, Ł. (2014) Lipoxygenase inhibitors and antioxidants from green coffee—mechanism of action in the light of potential bioaccessibility, *Food Research International,* 61, 48-55.

George, S.E., Ramalakshmi, K. and Rao, L.J.M. (2008) A Perception on Health Benefits of Coffee, *Critical Reviews in Food Science and Nutrition,* 48, 464-486.

Givens, D. I. and Barber, W. P. (1986) In vivo evaluation of spent coffee grounds as a ruminant feed, *Agricultural Wastes*, 18, 69-72.

Gómez-De La Cruz, F.J., Cruz-Peragón, F., Casanova-Peláez, P.J. and Palomar-Carnicero, J.M. (2015) A vital stage in the large-scale production of biofuels from spent coffee grounds: The drying kinetics, *Fuel Processing Technology,* 130, 188-196.

González, A.S., Plaza, M.G., Pis, J.J., Rubiera, F. and Pevida, C. (2013) Post-combustion CO_2 capture adsorbents from spent coffee grounds, *Energy Procedia*, 37, 134-141.

Hall, S., Desbrow, B., Anoopkumar-Dukie, S., Davey, A.K., Arora, D., McDermott, C., Schubert, M.M., Perkins, A.V., Kiefel, M.J. and Grant, G.D. (2015) A review of the bioactivity of coffee, caffeine and key coffee constituents on inflammatory responses linked to depression, *Food Research International*, 76, 626-636.

Hachicha, R., Rekik, O., Hachicha, S., Ferchichi, M., Woodward, S., Moncef, N., Cegarra, J. and Mechichi, T. (2012) Co-composting of spent coffee ground with olive mill wastewater sludge and poultry manure and effect of Trametes versicolor inoculation on the compost maturity, *Chemosphere,* 88, 677-682.

Hassan, F.A., Ismail, A., Hamid, A.A., Azlan, A. and Al-Sheraji, S.H. (2011). Characterisation of fibre-rich powder and antioxidant capacity of Mangifera pajang K. fruit peels. *Food Chemistry,* 126, 283-288.

Higdon, J.V. and Frei, B. (2006) Coffee and health: A review of recent human research. *Critical Reviews in Food Science and Nutrition,* 46, 101–123.

Horio, M., Suri, A., Asahara, J., Sagawa, S. and Aida, C. (2009) Development of Biomass Charcoal Combustion Heater for Household Utilization, *Industrial and Engineering Chemistry Research,* 48, 361-372.

Jiménez-Zamora, A., Pastoriza, S. and Rufián-Henares, J.A. (2015) Revalorization of coffee by-products. Prebiotic, antimicrobial and antioxidant properties, *LWT - Food Science and Technology,* 61, 12-18.

Kim, J.Y., Jung, K.S. and Jeong, H.G. (2004) Suppressive effects of the kahweol and cafestol on cyclooxygenase-2 expression in macrophages, *FEBS Letters*, 569, 321-326.

Kotowski, U., Heiduschka, G., Seemann, R., Eckl-Dorna, J., Schmid, R., Kranebitter, V., Stanisz, I., Brunner, M., Lill, C. and Thurnher, D. (2015)

Effect of the coffee ingredient cafestol on head and neck squamous cell carcinoma cell lines, *Strahlentherapie und Onkologie,* 191, 511-517.

Kwon, E.E., Yi, H. and Young Jae Jeon, Y.J. (2013) Sequential co-production of biodiesel and bioethanol with spent coffee grounds, *Bioresource Technology*, 136, 475-480.

Limousy, L., Jeguirim, M., Labbe, S., Balay, F. and Fossard, E. (2015) Performance and emissions characteristics of compressed spent coffee ground/wood chip logs in a residential stove, *Energy for Sustainable Development,* 28, 52-59.

Limousy, L., Jeguirim, M., Dutournié, P., Kraiem, N., Lajili, M. and Said, R. (2013) Gaseous products and particulate matter emissions of biomass residential boiler fired with spent coffee grounds pellets, *Fuel,* 107, 323-329.

Lucas, M., Mirzaei, F., Pan, A., Okereke, O.I., Willett, W.C., O'Reilly, E.J., Koenen, K. and Ascherio, A. (2011) Coffee, caffeine, and risk of depression among women, *Archives of Internal Medicine,* 171, 1571–1578.

Magalhães, L.M., Machado, S., Segundo, M.A., Lopes, J.A. and Páscoa, R.N.M.J. (2016) Rapid assessment of bioactive phenolics and methylxanthines in spent coffee grounds by FT-NIR spectroscopy, *Talanta*, 147 460-467.

Melo, J.C.S. (2013) Evaluation of the technological and antioxidant potential of coffee husks as substrates in the production of dietary fibers. M.Sc. Dissertation, Universidade Federal de Minas Gerais, Brazil (in portuguese).

Mikami, Y. and Yamazawa, T. (2015) Chlorogenic acid, a polyphenol in coffee, protects neurons against glutamate neurotoxicity, *Life Sciences,* 139, 69-74.

Monente, C., Bravo, J., Vitas, A.I., Arbillaga, L., Paz De Peña, M. and Cid, C. (2015a) Coffee and spent coffee extracts protect against cell mutagens and inhibit growth of food-borne pathogen microorganisms, *Journal of Functional Foods*, 12, 365-374.

Monente, C., Ludwig, I.A., Irigoyen, A., Paz De Peña, M. and Cid, C. (2015b) Assessment of total (free and bound) phenolic compounds in spent coffee extracts, *Journal of Agricultural and Food Chemistry*, *63*, 4327-4334.

Murthy, P.S.. and Naidu, M.M. (2012) Recovery of phenolic antioxidants and functional compounds from coffee industry by-products, *Food and Bioprocess Technology*, 5, 897–903.

Mussatto, S.I., Carneiro, L.M., Silva, J.P.A., Roberto, I.C. and Teixeira, J.A. (2011a) A study on chemical constituents and sugars extraction from spent coffee grounds, *Carbohydrate Polymers,* 83, 368-374.

Mussatto, S.I., Ballesteros, L.F., Martins, S. and Teixeira, J.A. (2011b) Extraction of antioxidant phenolic compounds from spent coffee grounds, *Separation and Purification Technology,* 83, 173-179.

Nkondjock, A. (2009) Coffee consumption and the risk of cancer: An overview. *Cancer Letters,* 277, 121–125.

Panusa, A., Zuorro, A., Lavecchia, R., Marrosu, G. and And Petrucci, R. (2013) Recovery of Natural Antioxidants from Spent Coffee Grounds, *Journal of Agricultural and Food Chemistry,* 61, 4162–4168.

Páscoa, R.N.M.J., Magalhães, L.M. and Lopes, J.A. (2013) FT-NIR spectroscopy as a tool for valorization of spent coffee grounds: Application to assessment of antioxidant properties, *Food Research International*, 51, 579-586.

Pavlović, M.D., Buntić, A.V., Šiler-Marinković, S.S. and Dimitrijević-Branković, S.I. (2013) Ethanol influenced fast microwave-assisted extraction for natural antioxidants obtaining from spent filter coffee, *Separation and Purification Technology,* 118, 503-510.

Plaza, M.G., González, A.S., Pevida, C., Pis, J.J. and Rubiera, F. (2012) Valorisation of spent coffee grounds as CO_2 adsorbents for postcombustion capture applications, *Applied Energy*, 99, 272-279.

Pujol, D., Liu, C., Gominho, J., Olivella, M.À., Fiol, N., Villaescusa, I. and H. Pereira, H. (2013) The chemical composition of exhausted coffee waste, *Industrial Crops and Products,* 50, 423-429.

Ramalakshmi, K., Rao, L, Takano-Ishikawa, Y and Goto, M. (2009) Bioactivities of low-grade green coffee and spent coffee in different in vitro model systems, *Food Chemistry,* 115: 79–85.

Reis, N., Franca, A.S. and Oliveira, L.S. (2013a) Performance of diffuse reflectance infrared Fourier transform spectroscopy and chemometrics for detection of multiple adulterants in roasted and ground coffee, *LWT - Food Science and Technology,* 53, 395-401.

Reis, N., Franca, A.S. and Oliveira, L.S. (2013b) Quantitative evaluation of multiple adulterants in roasted coffee by Diffuse Reflectance Infrared Fourier Transform Spectroscopy (DRIFTS) and chemometrics, *Talanta,* 115, 563-568.

Reyes-Caudillo, E., Tecante, A. and Valdivia-López, M.A. (2008) Dietary fibre content and antioxidant activity of phenolic compounds present in Mexican chia (Salvia hispanica L.) seeds, *Food Chemistry,* 107, 656- 663.

Rocha, M.V.P., Lima De Matos, L.J.B., Pinto De Lima, L., Figueiredo, P.M.S., Lucena, I.L., Fernandes, F.A.N. and Gonçalves, L.R.B. (2014) Ultrasound-assisted production of biodiesel and ethanol from spent coffee grounds, *Bioresource Technology*, 167, 343-348.

Rodríguez-Ambriz, S.L., Islas-Hernández, J.J., Agama-Acevedo, E., Tovar, J. and Bello-Pérez, L.A. (2008) Characterization of a fibre-rich powder prepared by liquefaction of unripe banana flour, *Food Chemistry*, 107, 1515-1521.

Sikka, S. S. and Chawla, J. S. (1986) Effect of feeding spent coffee grounds on the feedlot performance and carcass quality of fattening pigs, *Agricultural Wastes*, 18, *305-308.*

Sikka, S. S., Bakshi, M. P. S. and Ichhponani, J. S. (1985) Evaluation in vitro of spent coffee grounds as a livestock feed, *Agricultural Wastes*, 13, 315-317.

Silva, M.A., Nebra, S. A., Machado Silva, M. J. and Sanchez, C. G. (1998) The use of biomass residues in the Brazilian soluble coffee industry, *Biomass and Bioenergy*, 14, 457-467.

Simões, J., Nunes, F.M., Domingues, M.R. and Coimbra, M.A. (2013) Extractability and structure of spent coffee ground polysaccharides by roasting pre-treatments, *Carbohydrate Polymers*, 97 (2013), pp. 81–89.

Speer, K. and Kölling-Speer, I. (2006). The lipid fraction of the coffee bean. *Brazilian Journal of Plant Physiology*, *18*(1), 201-216.

Suzuki, A., Kagawa, D., Ochiai, R., Tokimitsu, I. and Saito, I. (2002). Green coffee bean extract and its metabolites have a hypotensive effect in spontaneously hypertensive rats. *Hypertensive Research*, 25, 99–107.

Tango, J. S. (1971) Utilização industrial do café e dos seus subprodutos (Industrial utilization of coffee and its byproducts) *Boletim do ITAL*, 28, 48-73 (in Portuguese).

Vardon, D.R., Moser, B.R., Zheng, W., Witkin, K., Evangelista, R.L., Strathmann, T.J., Rajagopalan, K. and Sharma, B.K. (2013) Complete Utilization of Spent Coffee Grounds To Produce Biodiesel, Bio-Oil, and Biochar, *ACS Sustainable Chemistry and Engineering*, *1*, 1286-1294.

Vilela, W.S. (2015) Characterization of dietary fibers from spent coffee grounds after treatment with hidrogen peroxide, M.Sc. Dissertation, Universidade Federal de Minas Gerais, Brazil (in portuguese).

Viuda-Martos, M., Ruiz-Navajas, Y., Martin-Sánchez, A., Sánchez-Zapata, E., Fernández-López, J., Sendra, E., Sayas-Barberá, E., Navarro, C. and Pérez-Álvarez, J.A. (2012). Chemical, physico-chemical and functional properties of pomegranate (Punica granatum L.) bagasses powder co-product, *Journal of Food Engineering,* 110, 220-224.

Xu, H., Wang, W., Liu, X., Yuan, F. and Gao, Y. (2015) Antioxidative phenolics obtained from spent coffee grounds (Coffea arabica L.) by subcritical water extraction, *Industrial Crops and Products,* 76, 946-954.

Xu, G., Murakami, T., Suda, T., Matsuzawa, Y. and Tani, H. (2006) Gasification of Coffee Grounds in Dual Fluidized Bed: Performance Evaluation and Parametric Investigation, *Energy Fuels*, 20, 2695– 2704.

Wei, F. and and Tanokura, M. (2015) Chemical Changes in the Components of Coffee Beans during Roasting. In V.R. Preedy (Ed.), Coffee in Health and Disease Prevention (pp. 83-91), San Diego: Academic Press.

Zuorro, A. (2015) Optimization of polyphenol recovery from espresso coffee residues using factorial design and response surface methodology, *Separation and Purification Technology,* 152, 64-69.

Zuorro, A. and Lavecchia, R. (2012) Spent coffee grounds as a valuable source of phenolic compounds and bioenergy, *Journal of Cleaner Production,* 34, 49-56.

In: Food Waste
Editor: Garrett Leonard Riley

ISBN: 978-1-63485-025-4

Chapter 4

POTENTIAL APPLICATIONS OF FOOD WASTES TO BIOENERGY BY ANAEROBIC PROCESSES

Sandra I. Maintinguer[1,2,*] and Lorena Oliveira Pires[3]
[1]UNESP – São Paulo State University, Bioenergy Research Institute (IPBEN), Rio Claro, SP, Brazil
[2]UNIARA - University Center of Araraquara, Post-Graduation on Regional Development and Environment - Araraquara, SP, Brazil
[3]UNESP – São Paulo State University, Chemistry Institute, Araraquara, SP, Brazil

ABSTRACT

Food wastes have increased worldwide due to the growth of economy and population of the world. They are considered resources renewable, sustainable, and easily available from restaurants, kitchens, food processing plants, etc. and they are mainly composed of rice, vegetable, and meat, which could be easily degraded by microorganisms.

Feasible raw materials from food wastes are generated largely and they are disposed in municipal landfill or dumping sites causing environmental problems. The high moisture and organic content from these wastes can be treated in anaerobic digestion because it provides volume and mass reduction up to 50% from the input material. The bio-

* Corresponding author: mainting2008@gmail.com.

products generated during the anaerobic biologic processes could be converted to energy of high value-added including methane, hydrogen, alcohols, enzymes, organic acids, biopolymers, bio plastics, among others.

The anaerobic processes provide a waste management option for food wastes, while offers the opportunity for recovering marketable products in the form of biogas as methane and hydrogen that it has been considered attractive and potential method to obtain renewable bio-fuels. Additionally, they can be sold to a local utility, especially in the developing countries or rural areas, where energy may be not sufficient.

Therefore, this chapter will contribute to innovation in researches on the valorization techniques that have been developed for the use of food wastes and the future directions for more effective ways of applications these residues, in a sustainable manner, thus enabling cost reduction and clean energy generation.

Keywords: anaerobic digestion, food waste, hydrogen, methane, ethanol

INTRODUCTION

The major quantities of municipal solid waste is compound by food waste disposed at landfills as municipal solid wastes. Its degradation creating environmental problems as odor, air pollution, emissions of carbon dioxide and methane, among others toxic gases. According to Food and Agricultural Organization of the United Nations it is discarded 1.3 billion tons of food waste without any further use (Karmee, 2016, 945, 947). Big spaces are required to landfills implantation in the cities with complex structures and high costs are an important consequence. In this sense, it is necessary to solve the problems caused because of generation of food waste.

There are some possible technologies to destination of food wastes and thus does not contaminate the environment. Two classic examples are anaerobic digestion and incineration.

According to Pham et al. (2015, 399, 407), food waste is a solid residue with great potential for generating energy. There are diverse technologies that have been studied such as biological and thermochemical incineration, pyrolysis, gasification and hydrothermal oxidation (Figure 1).

Nevertheless, during the process of anaerobic digestion high amounts of biogas are generating, mainly methane and hydrogen. It can be an alternative way to use food wastes to produce biofuels. Methane generation is currently developed at industrial level for full-scale application. However, the hydrogen biological production from food waste is widely studied but, nowadays, no full-scale applications have been used (Girotto et al., 2015, 32). In addition, lipids and carbohydrates present on food waste composition can be converted to biodiesel, bio-oil and bio-ethanol via anaerobic digestion. This process can be advantageous and according to Murphy et al. (2004, 407, 415), 1 m^3 of biogas generated equals to 21 MJ of energy during the anaerobic digestion. This energy could be converted in 2.04 kWh of electricity with an efficiency generation of 35%.

Arvanitoyannis and Varzakas (2009, 57, 125) listed the main advantages of the biodegradation waste management as follows: It allows reducing the volume of organic wastes; the biological hazard of the wastes can be controlled; this system may be compatible with the other biological ELSS (greenhouses); the biogas manufactured can be used to produce electricity; the water obtained in the biodegradation processes may be used for the other needs of the space vehicle; a valuable effluent is also obtained, which eventually can be used as an excellent soil conditioner after minor treatments and high OLRs (Organic Load Rates) and low sludge production are among the many advantages anaerobic process exhibit over other biological unit operations.

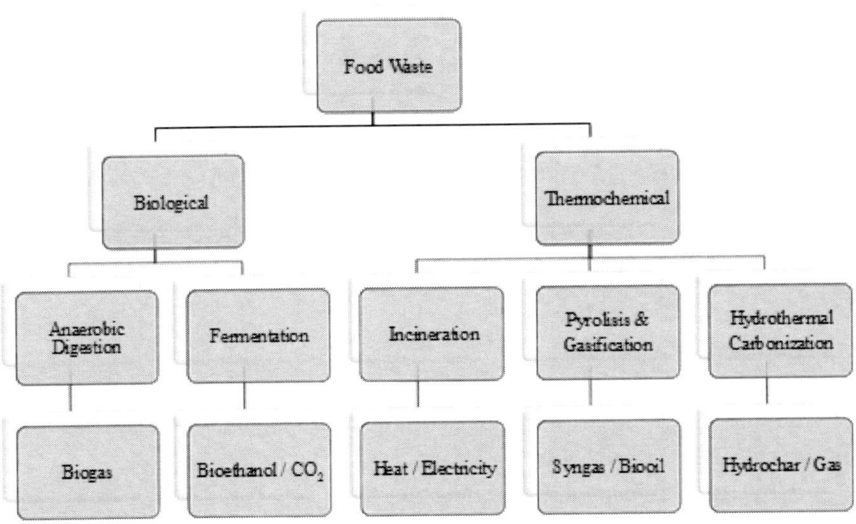

Figure 1. Summary of food-waste-to-energy technologies.

As the solid wastes produced from traditional market are tremendously nuisance to the environment, the use of food waste for biofuels and bioproducts by biologic processes could benefit the environment due to reduced amounts of food waste disposed of in landfills. In this sense, this paper proposes the anaerobic digestion for treating solid organic wastes that contained mainly vegetable waste with recovering bioenergy during the process.

Food Wastes as a Substrate in Anaerobic Digestion Process

Food wastes are generated largely and this residue includes fruits, vegetables, fish and poultry organs, intestine, meat, fats, sauces, condiments, soup pulp, herbal medicinal, pulp eggs, cheeses, ice cream, yogurts, coffee, tea, bread, cakes, biscuits, desserts, pet foods, among others. (Sitorus et al., 2013, 176, 181). Although the rich composition of the food wastes, they are discarded because they are considered resources without commercial value. However, it is important to know that the chemistry composition of food wastes include lipid, carbohydrate, amino acid, phosphate, vitamins and other carbon substances (Karmee, 2016, 945, 952). They also content sugar and protein in the range of 35.5 to 69% and 3.9 to 21.9%, respectively (Kiran et al. 2014, 389, 390).

An alternative to the food wastes is treating it biologically via anaerobic digestion. Anaerobic digestion is important because the biogas produced, rich in methane is suitable for energy production and it can replace fossil fuels.

Anaerobic Digestion is a process in which specific groups of anaerobic bacteria break down biodegradable materials in the absence of oxygen. The process is very efficient because it provides volume and mass reduction up to 50% from the input material during the biologic treatments of wastewater sludges, industrial and farm wastes (Sitorus et al. 2013, 180), for example.

The anaerobic digestion includes four phases: hydrolysis, acidogenesis, acetogenesis and methanogenesis (Figure 2).

Firstly, high molecular materials are hydrolyzed and break down into small organic substrates as fatty acids, glucose and aminoacids by exo-enzymes that are excreted by fermentative bacteria. Generally, the hydrolysis limits the rate of the overall anaerobic digestion process because the hydrolytic enzymes should be primarily adsorbed on the surface of solid substrates as showed in equation 1 (Kondusamy and Kalamdhad, 2014, 1823).

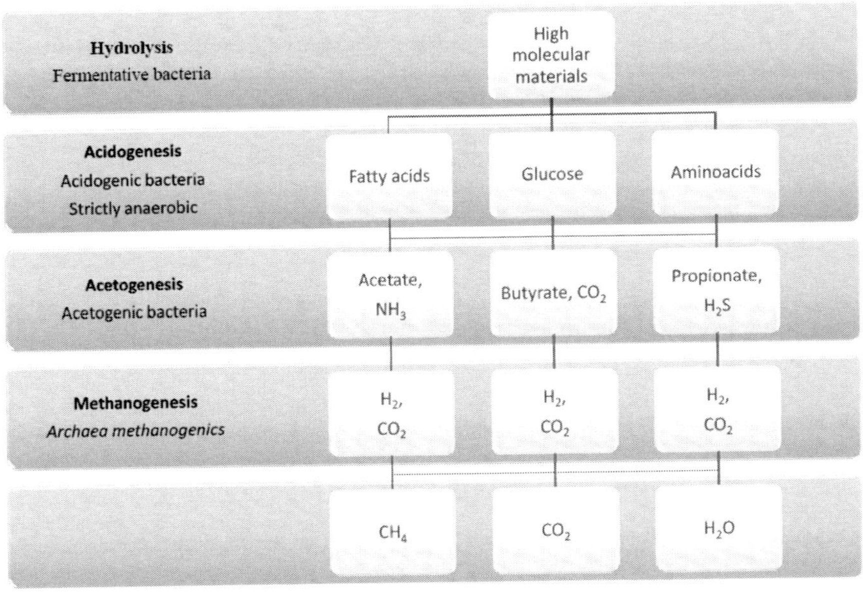

Figure 2. Anaerobic Digestion Scheme.

$$C_6H_{10}O_5 + H_2O \xrightarrow{hydrolysis} C_6H_{12}O_6 \qquad (1)$$

Many fermentative bacteria can participate in the hydrolysis process, such as *Bacteroides*, *Clostridium*, *Enterobacter*, *Lactobacillus*, *among others* (Gerardi, 2006).

The next phase of the anaerobic digestion is the acidogenesis, were the hydrolyzed materials are degraded into small organic substrates as volatile fatty acids (acetate, butyrate and propionate) with the generation of by-products such as NH_3, CO_2 and H_2S by acidogenic bacteria (equation 2). A very diverse group of bacteria is present at this phase and the most of them are microorganisms strictly anaerobic. The presence of oxygen utilizing bacteria is important to remove all the oxygen that might be introduced into the system. The acidogenic bacteria are able to metabolize organic material at low pH as near to 4.

$$C_6H_{12}O_6 \xrightarrow{\text{acid forming bacteria}} 3CH_3COOH \tag{2}$$

In the third phase, called acetogenesis, the organic substrates produced in the second phase are digested into acetate, H_2 and CO_2, which could be used by archaea methanogenics for the formation of methane gas. Ordinarily, the substrates which could be utilized by these specific microorganisms are: short-chain fatty acid (C1 – C6), alcohols (ethanol, buthanol) and gases (CO, CO_2 and H_2).

In the last step, CH_4 can be produced in two ways:

1. Fermenting acetic acid to CH_4 and CO_2 by acetoclastic methanogens (equation 3). In this case CO_2 is used as a source of carbon and hydrogen is used as a reducing agent by hydrogenotrophic methanogens;
2. By other bacterial species as given in reactions showed at equations 4 to 6.

$$CH_3COOH \xrightarrow{\text{methane forming bacteria}} CH_4 + CO_2 \tag{3}$$

$$CO_2 + 4H_2 \xrightarrow{\text{reduction}} CH_4 + 3H_20 \tag{4}$$

$$CO_2 + H_2O \xrightarrow{\text{hydrolysis}} H_3CO_3 \tag{5}$$

$$4H_2 + H_2CO_3 \xrightarrow{\text{reduction}} CH_4 + 3H_2O \tag{6}$$

The most microorganisms commonly found in the biogas reactors are archaea methanogenic, mainly *Methanobacterium*, *Methanothermobacter*, *Methanobrevibacter*, *Methanosarcina* and *Methanosaeta* (Kondusamy and Kalamdhad, 2014, 1822).

Among the advantages of anaerobic digestion are included the production of a renewable energy carrier, the possibility of nutrient recycling with reduction of waste volumes and generation of methane as a potential energetic gas.

Food wastes have been treated both separately and in co-digestion processes. The co-digestion is better for the digestibility and it can be achieved by mixing solid waste with diluted waste. Furthermore, the addition of different wastes during the anaerobic digestion of the food waste results in a better digestion performance by improving the content of the nutrients and even reducing the negative effect of toxic compounds (Kosseva, 2011). Food waste could be digested anaerobically and it would have the potential to generate 367 m^3 of biogas per dry ton at about 65% CH_4 with an energy content of 6.025 x 10^{-9} TWh/m^3 (Kondusamy and Kalamdhad, 2014, 1828, 1829).

The ideal conditions to apply the Anaerobic Digestion process in the treatment of food wastes depends on several parameters such as Temperature, Volatile Fatty Acids concentration, pH, reactors configurations, among others (Zhang et al., 2014, 383-392).

Ideal Parameters to Anaerobic Digestion of Food Wastes

Ideal environmental conditions are necessary to reach the best metabolic activities during the anaerobic digestion. Archaea methanogenics are microorganisms very sensible to adverse environmental mainly temperature and pH (Speece, 1996, 394, 399). So, it is very important to maintain optimal conditions for these microorganisms development. The rate of biogas production depends on several parameters, such as the nature of the substrate, temperature, pH, toxicity, mixing, nutrients, slurry concentration, retention time, digester type, carbon nitrogen ratio, alkalinity, initial feeding, total VS (Volatile Solids), COD (carbon Organic Demand), etc. (Kondusamy and Kalamdhad, 2014, 1825).

Temperature

The anaerobic bacteria can grow at three different conditions of temperature: psychrophilic (10-30°C), mesophilic (30-40°C) and thermophilic (50-60°C) (Zhang et al., 2014, 383-392). Temperature can influence directly the enzymes and co-enzymes activities, the methane yields, the effluent quality, ionization equilibrium, substrate solubility, fats and iron bioavailability during the anaerobic digestion process (Speece, 1996, 394-400). Higher temperature can inhibit the activity of hydrogenotropic methanogens in anaerobic process and to favor both hydrogen producing bacteria and spore forming bacteria (Maintinguer et al., 2008, 4309, 4312).

According to Kosseva (2011), the temperature variations during the anaerobic digestion are strictly dependent on the climatic conditions. In Sweden, the researches could obtain methane generations at psychrophilic conditions. In Europe, anaerobic digestion had been well established under thermophilic conditions for the treatment of the organic fraction of municipal solid waste. However, in the United States, thermophilic condition has been abandoned. In tropical countries, such as Tunisia and Brazil, where the ambient temperature is higher than 25°C during a period of more than 8 months in a year, mesophilic anaerobic digestion is readily applicable (Maintinguer et al., 2015b, 3180, 3188).

However, different temperatures condition can be tested in controlled systems as anaerobic reactors. Bouallagui et al., (2003, 85, 87) compared the anaerobic digestion performance in a tubular anaerobic digester on a laboratory scale. They compared the energy balance of the process under 55, 20 and 35ºC for several retention times (10 to 20 days) and feed concentration. The net energy production in the thermophilic digester was 195.7 and 49.07 kJ d^{-1} higher than that for the psychrophilic and mesophilic digesters, respectively. The relation between the daily production of biogas and the temperature indicates that for the same produced quantity of biogas, the size of the thermophilic digester can be reduced with regard to that of the psychrophilic and the mesophilic digesters (Kosseva, 2009, 57, 100).

Volatile Fatty Acids

The concentrations of volatile fatty acids (VFA) such as acetic, propionic, butyric and valeric acids can have effects in all phases of anaerobic digestion. According to Zhang et al. (2014, 383-392), VFA can accumulate during the anaerobic digestion process causing pH decrease and consequent fault of the biogas generation. Frequently, a mix of acetic and propionic acids are dominant in biogas production but, if the ratio between them were 1.4, can occur the anaerobic digestion failure. Methods for monitoring of VFAs such as gas-cromatography (GC) and high performance liquid chromatography (HPLC) are frequently used to analyze the anaerobic reactors performance. The information about the VFAs gives the status of the anaerobic digestion and it is also an important indicator for monitoring the process (Adorno et al., 2014, 406, 412).

pH

The pH is one of the most important parameters that can affect the efficiency of anaerobic digestion. Anaerobic bacteria need a specific range of pH for their development. Fermentative bacteria can survive in a pH range from 4.0 to 8.5 (Zhang et al., 2014, 383-392). However, archaea methanogenic can grow in a limiting range from 6.5 to 8.2 (Speece, 1996, 394-400). It has been proven that the optimal pH range to obtaining maximal biogas yield in anaerobic digestion is from 6.5 to 7.5. Moreover, acetic and butyric acids are dominant at low pH while acetic and propionic acids are present at pH 8.0 (Zhang et al., 2014, 383-392, Maintinguer et al., 2015b, 3180, 3181). It is important to know that pH can vary with substrate, bicarbonate concentration, alkalinity of the system and CO_2 among other factors. Additions of $NaHCO_3$ and NaOH are made for pH controlling during anaerobic digestion process (Kondusamy and Kalamdhad, 2014, 1826).

C/N Ratio

Carbon and Nitrogen are elements necessary as nutrient of anaerobic bacteria. Carbon is used for energy and nitrogen is used to form proteins in the cell (Jain et al., 2015, 142, 145). The bacterial populations decrease if the concentrations of nitrogen were low and it might take longer duration to digest the available carbon. Anaerobic bacteria can use carbon 25 to 35 times higher than nitrogen, during anaerobic digestion (Kondusamy and Kalamdhad, 2014, 1829). According to Zhang et al. (2014, 383-392) C/N ratio range of 20 to 30 was considered to be the optimal to anaerobic digestion. The ideal C/N ratio varies with the type of substrate to be degraded. However, inadequate C/N ratios could result in high ammonia formation and/or high volatile fatty acids (VFA) accumulation during the anaerobic digestion process. The effect of the ammonia inhibition can be solved by dilution with water; it decreases the toxicity in anaerobic digestion process. More nitrogen addition leads to ammonia formation while its absence leads to nitrogen deficiency (Kondusamy and Kalamdhad, 2014, 1824). Therefore, C/N ratio range must be adequate because both ammonification and VFAs are important inhibitors for the methanogenic activity and they cause possible failure of the anaerobic digestion process with no methane generation (Jain et al., 2015, 142, 147). Thus, carbon and nitrogen control the bacterial population and the optimum C/N ratio can depend on them during the anaerobic process.

Nutrients

C, H_2, O_2, N_2, P and S are the major nutrients required for the anaerobic bacteria during the anaerobic digestion process (Jain et al., 2015, 142, 145). The addition of trace elements such as Ni, Cr, Cu, Co, Zn, N, K, Mg, Ca, Al is necessary to start the anaerobic digestion (Kondusamy and Kalamdhad, 2014, 1822, Zhang et al., 2014, 383-392). These elements are required both for enzimatic synthesis and for maintain their activities on anaerobic bacteria. However, the wastes characteristics can present others essential nutrients such as N_2 and P. These compounds are always founded in short supply and they maintain the nutrients balance to obtain maximum biogas production (Jain et al., 2015, 142, 150). Inhibition effects of anaerobic bacteria may occur if the concentrations of these elements were too high. Some elements such as heavy metals are not biodegradable and they can accumulate causing inhibition by enzyme and structure disruption of anaerobic bacteria. Many others factors with metal concentrations may cause inhibition such as chemical forms of the metal, pH and redox potential (Zhang et al. 2014, 383-392).

Reactors Configurations

In order to control the anaerobic process and improve the biogas production from food waste, several researches have been developed with different reactors configurations such as single or two stage reactor, upflow anaerobic solid state reactor, solid state anaerobic digester and hybrid reactors for the treatment of organic waste.

When all anaerobic digestion, namely hydrolysis, acidogenesis, acetogenesis and methanogenesis, occurs in a single vase or reactor, it can be observed the conversion of organic compounds to CH_4, H_2S, NH_3 and CO_2 and this equipment is named single stage reactor. However, when the hydrolysis and acidogenesis take place in initial reactor and the utilization of those acids by methanogenesis is anticipated to take place in the final reactor, the system is named two-stage reactor (Kondusamy and Kalamdhad, 2014, 1826). Previews researches such as Mata-Alvarez (1987, 844, 846), Lissens, et al. (2001, 91, 94) and Battimelli et al. (2009, 6395, 6398), demonstrated that the performance of two-stage anaerobic reactors is more efficient that the single stage reactor. The separation of digesters as hydrolyzers and methanizers changes the process dynamics of the digesters by allowing the individual bacterial species separately to perform the hydrolysis and methanogenesis

(Kondusamy and Kalamdhad, 2014, 1826). When the system is operated as a two-stage digester, the slow growing methanogens is not affect by variations in environmental conditions such as pH, or organic concentration Kim, et al., (2008, 4394, 4396).

In order to compare two methane fermentation systems, Park et al. (2008, 48, 50) operated and fed with artificial kitchen waste a lab-scale single-phase and a two-phase thermophilic reactor. After analyze different hydraulic retention time (HRT) and organic load rate (OLR) in both reactors the authors concluded that even though the two-phase reactor is more stable and easier to maintain, efficient kitchen waste degradation can be accomplished in both systems with highest methane recovery ratio of 90%.

Kim et al. (2006, 328, 330) operated a three-stage methane fermentation system developed to digest food waste. The three stages were: semi-anaerobic hydrolysis, anaerobic acidogenesis and strictly anaerobic methanogenesis. Several temperatures and hydraulic retention time (HRT) were used and, with liquor food waste, thermophilic digester showed a higher rate of Chemical Oxygen Demand (COD) removal than mesophilic digesters. About the rates of biogas and methane production, thermophilic condition also was better than mesophilic.

Several different configurations that have been used in anaerobic treatments of wastes including Continuously Stirred Tank Reactor (CSTR), Fluidized Bed Reactor (FBR), Anaerobic Sequencing Batch Reactors (AnSBR) and Upflow Anaerobic Sludge Blanket (UASB) reactors (Kiran et al., 2014).

One of the most pronounced development in anaerobic treatment technology is the Upflow Anaerobic Sludge Blanket (UASB) reactor. The feasibility of UASB reactors concept for treating soluble wastewaters has been sufficiently demonstrated at pilot and full scale (Latif et al., 2012, 114, 117). This reactor configuration presents satisfactory results in the wastewater treatment such as domestic sludge (Seghezzo, et al., 1998, 175, 176), vinasse (Ferraz, et al., 2016, 245), surfactant (Okada et al., 2013, 125), and others. Latif et al. (2012, 114, 115) operated a UASB reactor to explore the feasibility of mesophilic and thermophilic anaerobic digestion of liquid food waste at various organic loading rates and temperatures for waste reduction and biogas production. At this work, maximum COD removal efficiency was 93.67%; biogas and methane yield were recorded to a maximum of 1.364 L/g-COD removed.d, 0.912 L/g-COD removed.d, and average methane content of biogas was 63%. The food waste, consisting in a variety of carbon sources such as rice, coconut, vegetables, fruit, chicken meat, fish, and eggshells and the authors concluded that the treatment process can also be extendable for more than 10 weeks without any measurable problem.

In the last decades, Anaerobic Sequencing Batch Reactors (AnSBR) have been researched and considered a potential alternative for treating several types of wastewater because they promoted good solids retention in the system and improved process control (Dague et al., 1992, 2429; Contrera et al., 2014, 385, 386). The main advantages of AnSBR are the possibility of achieving high solids retention, high organic matter removal efficiency, providing better effluent quality control and the possibility of suitable process control (Dague et al., 1992, 2429; Oliveira et al., 2010, 2499, 2500). About food wastes, this reactor configuration was tested by Kim et al. (2004, 1607, 1610) as anaerobic co-digestion of food waste and sewage sludge for hydrogen production. The authors observed that the specific hydrogen production potential of food waste was higher than that of sewage sludge. However, hydrogen production potential increased as sewage sludge composition increased up to 13–19%. So, the authors concluded that food waste and sewage sludge were considered as a suitable main substrate and a useful auxiliary substrate, respectively, for hydrogen production. The metabolic results indicated that the fermentation of organic matters was successfully achieved.

Table 1. Optimum conditions for Anaerobic Digestion of food wastes

Parameters	Optimum conditions	References
Temperature	Mesophilic (25°C – 35°C) Thermophilic (55°C)	Boulagui et al. (2003)
Volatile Fatty Acids	Mix acetic and propionic	Zhang et al. (2014)
pH	6.5 – 8.2 4.0 – 8.5	Speece (1996) Zhang et al. (2014)
Nutrients	C, H_2, N_2, O_2, P, S Ni, Cr, Cu, Co, Zn, N, K, Mg, Ca, Al	Jain et al. (2015) Zhang et al. (2014)
C/N ratio	20-35 20-30	Kondusamy and Kalamdhad (2014) Zhang et al. (2014)
Reactors Configurations	Single-stage Two-stage Tree-stage UASB AnSBR	Kondusamy and Kalamdhad (2014) Battimelli (2009) Kim et al. (2006) Lafit et al. (2012) Kim et al. (2004)

Kim et al. (2010, 1590, 1591) studied a pilot scale anaerobic sequencing batch reactor (ASBR) treating food waste collected from a school cafeteria and supplemented with steamed rice and peptone intending to vary the C/N ratio. In order to improve the reactor performance, alkaline shock (pH 12.5 for 1 day) was imposed on the entire mixed liquor. This strategy was efficient and the H_2 yield increased to over 0.9 mol H_2/mol hexose. The authors also studied the layers formed into the reactor and the VSS (Volatile Suspense Solids) concentrations of each layer suggest that decanting the upper layer might play an important role in improving the treatment.

The optimum conditions for anaerobic digestion of food wastes can be summarized according the operational parameters as showed in Table 1. In addition, the anaerobic processes depend on the composition of food wastes and the inoculum used to maintain high efficiencies on biogas production. The age and the storage of the food wastes are others conditions that can affect the anaerobic digestion (Krishna and Kalamdhad, 2014). So, the applications of food wastes in anaerobic processes to valuable biogas need to be focused for the best biochemical methane and or/hydrogen potential and they would be research case to case.

ENERGY POTENTIAL FROM FOOD WASTE ON ANAEROBIC PROCESSES

The food wastes utilization by anaerobic process to obtaining bio-products such as biogas can decrease their accumulation in landfills. Due to its organic- and nutrient-rich composition, food waste can be utilized as a useful resource for production of biofuel through various fermentation processes. So far, valorization of food waste has attracted increasing interest, with biogas, hydrogen, ethanol and biodiesel as final products. Due to inherent chemical complexity, FW also can be utilized for production of high-value materials, such as organic acids, biodegradable plastics and enzymes. However, it should be noted that the market demand for such chemicals is much smaller than that for biofuels (Kiran, et al., 2014, 389, 390). According to Girotto et al. (2015, 35) two anaerobic processes can be convert food wastes into biofuels or energy: 1) anaerobic digestion to produce methane and 2) dark fermentation to produce hydrogen.

There are researches with anaerobic reactors applied to food wastes utilization by productions of methane and hydrogen generation. Some parameters are important to optimize the ideal conditions of anaerobic processes such as anaerobic reactors configurations, inoculum sources, pre-treatments of food wastes, operation time, feedstock characteristics, physical and chemical characteristics of the food wastes (volatile solid and nutrient contents, particle size), among others. These parameters can affect the biogas production. However, successful experiences have been made about both hydrogen and methane generations by anaerobic processes from food wastes re-utilization.

Methane Production

Methane generation by anaerobic digestion is an acceptable solution for food waste because of its low cost and its utilization as a renewable energy source. In addition, the nutrient-rich degraded produced can also be used as fertilizer or soil conditioner. Methane gas can be burned to produce both heat and electricity using internal combustion engines or micro turbines in a cogeneration arrangement where the electricity and waste heat generated are used to warm the digesters or to heat buildings (Sitorus et al., 2013, 179). Any excess electricity can be sold to suppliers or put into the local grid. According to Girotto et al. (2015, 38), 1.32 x 10^9 m^3 methane can be produced annually and may be generate 2.6 x 10^7 GJ of energy using the total food waste generated in the world.

High efficiencies of anaerobic digestion of food wastes to produce methane have been obtained in anaerobic processes. Kim et al. (2006, 328, 330) inoculated semi-continuous reactors with anaerobic bacteria isolated from landfill soil during 30 days and they observed efficiencies of 64.5% of methane generation in the biogas generated from a Korean liquor waste food. Banks et al. (2011, 612, 615) tested bioreactors receiving food waste collected mainly from domestic kitchens during 200 days with 15 kg COD/m^3 d inoculated with seed sludge and they obtained 90% of methane into biogas generations. Park et al. (2008, 53, 54) tested UASB reactor inoculated with sludges from various sources and feed with artificial kitchen waste with 54 Kg VS/m^3.d and they observed generations of biogas containing 68% of methane. The simulated kitchen waste was composed of vegetables, fruits, rice, noodle and bread, meat, fish and eggs. Latif et al. (2012, 120, 121) treated food waste collected from the canteen and the residential areas of University Malaysia Pahang operating a hydrolytic reactor methanogenic during 19 days with seed sludge, 10 kg SV/m^3 d and they verified variations of methane into biogas of

63-70%. Kim et al. (2008, 4394, 4396) inoculated anaerobic bacteria from a mix of landfill soil and cow manure in anaerobic batch reactor during 60 days of operation and they obtained 220 mL CH_4/g VS. The food waste to be treated was collected from a cafeteria at a Korean university and was crushed into small particles from 3 to 5 mm.

Hydrogen Production

The hydrogen has a high-energy yield (142.35 kJ/g) when used as compressed gas (Kiran et al., 2014, 391). H_2 production can be obtained by dark fermentation with food wastes rich in carbohydrates. In addition, the composition of food wastes, pre-treatments and process configurations may affect H_2 production (Kiran et al., 2014, 389, 393). Mixed cultures have been applied for H_2 production from waste materials (Maintinguer et al., 2011, 13508, 3511). However, in the waste materials there are hydrogen consuming bacteria such as archaea metanogenic (Kiran et al., 2014, 396, 397). Hydrogen consuming bacteria such as lactic acid bacteria are the most abundant species in untreated food wastes, while H_2-producing bacteria are dominant in the pre-treated food wastes (Kiran et al., 2014, 397). Therefore, the seed biomass from food wastes must be pretreated. Some treatments such as heat, acid or alkaline are efficient to inactivate hydrogenotrophic microorganisms and select hydrogen producing bacteria such as *Clostridium* and *Enterobacter* species (Maintinguer et al., 2008, 4309, 4315). However, pretreatment can be increase costs in large scale operations, but the technological and economic challenges associated with the fermentative H_2 production and its purification, storage, and distribution may also be improved to accelerate the application of bio H_2 as bioenergy (Maintinguer et al., 2015b, 3180, 3185).

There were observed hydrogen generations by dark fermentation during operation of anaerobic reactors fed with food wastes. Shin et al. (2004, 1355, 1362, 1363) worked with a milled food waste collected from a dining hall (without animal ones and clamshells) operating anaerobic batch reactors inoculated with acidogenic culture during 6 days and they obtained yields of 1.8 mol H_2/mol hexose. Kim et al. (2004, 1607, 1609, 1616) applied anaerobic batch reactor with heat chock seed sludge to do co-digestion of food waste and sewage sludge and they observed yields of 0.9 mol H_2/mol hexose. Lee et al. (2010, 13458, 13462) worked with a Sequencing Batch Reactor (SBR) during long time of operation (96 days) and they obtained yields of 2.5 mol H_2/mol hexose from a vegetable kitchen waste (e.g., cabbage, water celery, cauliflower, etc.). Kim et al. (2010, 1590, 1592, 1594) studied the hydrogen potential in a pilot scale Anaerobic Sequencing Batch Reactor (ASBR) with application of alkaline pretreatment (pH 12.5 during 1 day) in heat shocked seed sludge with yields of 0.9 mol H_2/mol hexose. In this work, the food waste was collected from a school cafeteria but steamed rice and peptone were intentionally added to control the C/N ratio. In another study, food waste collected from a dining hall was ground after removal of animal bones and clamshells by Youn et al. (2005, 32-35). They obtained hydrogen generations with anaerobic batch reactors operated during 5 days with anaerobic seed sludge and yields of 2.4 mol H_2/mol hexose.

Others Bio-Products

Some bio-products may be obtained from food wastes such as biogas, hydrogen, ethanol, organic acids and biodiesel. Ethanol is the main bio-product that could be used as biofuel.

Diverse food wastes have been utilized for the production of bio-ethanol such as agro-industrial wastes (banana peel, sugar beet pulp, pineapple, grape pomace, potato peel, citrus), cafeteria food waste and household food waste (Pham et al., 2015, 402). Traditionally, bioethanol can be produced from cellulose, potato, rice, and sugar cane. The starch present in these crops can be easily converted to glucose by commercial enzymes and subsequently fermented to ethanol. As the hydrolysis of cellulose is more difficult, the use of abundant and cheap wastes such as lignocellulosic, municipal and food wastes has been explored as alternative substrates for ethanol production (Kiran, et al., 2014, 394).

The industries that transform food wastes in energy can be called Biorefineries. According to the International Energy Agency (IEA), biorefinery is a way for sustainable biomass processing in a wide range of bio-products including food wastes in bioenergy products (biofuels, electricity and heat) (Biernat and Grzelak, 2015, 436-438). They can convert organic wastes into a variety of products as bio-based fuels. This kind of refinery, convert corn, sugar cane and other plant based material even as petroleum refineries convert oil into fuels and ingredients for use in a wide variety of consumer products.

There are several types of food wastes and their production location is an important factor for their utilization as base to energy production. For example, food wastes produced from agriculture and food processing are abundant and concentrated in specific location, therefore, those materials could be less susceptible to deterioration than food wastes produced at household level (Galanakis, 2012, 68, 72), and they can be used as inputs for industries of energy sector. Availability of food wastes and location of potential users define the feasibility of industrial symbiosis (Mirabella et al., 2014, 28, 36).

Valorization routes of food waste in biorefinery chains include both extraction of high-value components already present in the substrates to be used for nutrition or pharmaceutical applications and conversion into chemicals, materials or biofuels by the use of chemical or biological processes. Type, origin, seasonal generation and territorial distribution of FW will affect transport logistic for its utilization and compatibility with the transformation process. For these reasons, large fluxes of agro-industrial wastes seem to be more suitable for biorefinery chains than to anaerobic digestion processes. Bioerefineries are required for extraction or production of specific commodities with stability of supply and substrate homogeneity while waste waters from household or restaurants would be more indicated for biologic treatment processes where composition variability, origins and contaminations do not represent limits for the selected process (Pfaltzgraff ad Clark, 2014, 3-33). According to Sen et al. (2016, 547, 549), biorefinery would be viable if the costs of obtain of bio-products from FW are effective. In addition, legislations have to be created to guarantee support of bio-products obtained from food waste and its application for a biorefinery concept to be implemented (Girotto et al., 2015, 40).

A scheme with the possible products obtained from waste food to bioenergy in a Biorefinery chain is proposed (Figure 3) by anaerobic processes. In this case, food waste is the main substrate for bioenergy generation. It can feed anaerobic reactors in two phases. The first anaerobic reactor can degrade food waste in hydrogen, organic acids and alcohols by dark fermentation. In the second phase, organic acids can feed anaerobic reactors able to methane generation. Bio-products obtained such as hydrogen, methane, bio-ethanol and organic acids could be used to fuel cell, heating and industries (food, cosmetics and pharmaceutical), respectively. In addition, the use of food waste to bioenergy could need pre-treatments for commercial scale (Sen et al., 2016) and the biomass producing during the processes as sludge can be used as organic fertilizers. In this sense, the use of food waste as a feedstock employing the dark fermentation and anaerobic digestion processes is potential and sustainable.

Despite it being possible the food wastes conversions in interesting resources as energy source, the ideal situation would be the application of all the 5R on this kind of residue. The 5 Rs are: **R**eduction at source, **R**euse, **R**ecycle, **R**ecovery and **R**efinery (Dung et al., 2014, 307, 310). Reduction implies in avoid the waste preparing just enough food and not over-preparation. This way, the intention is consuming all the food prepared. Reuse means store the spare food and consume it later. Another option is donating the overeating. Recycle is to use the food waste to generate something useful such as energy (as discussed in this text). Another example is using food waste as base to animal food. Recovery is the entire process involving food waste, namely, collect, sort, process and conversion in a product value. Finally, Refinery is the industry capable of to transform food waste in hydrogen, methane and others products with high values by biologic processes as dark fermentation and anaerobic digestion.

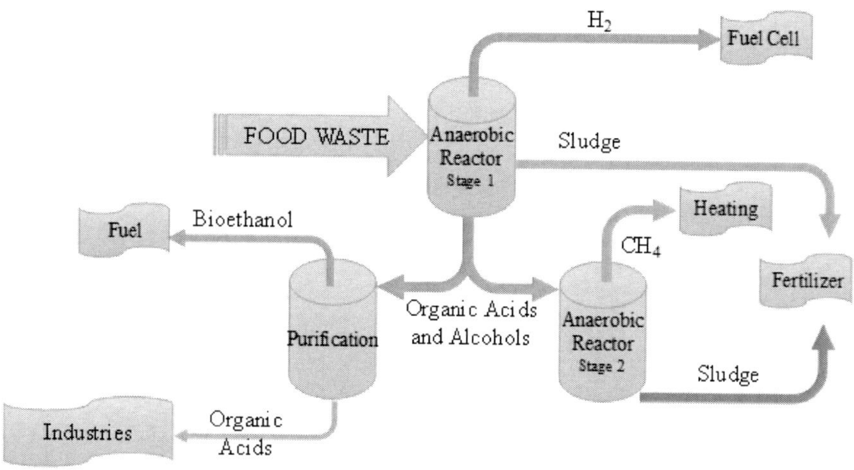

Figure 3. Scheme of food waste to energy in a Biorefinery chain.

Conclusion

After all that has been discusses, it can be concluded that the rich components of organic matter from food waste are suitable to be treated by anaerobic digestion, which would be reduce the volumes of waste in landfills.

So, developments in anaerobic processes provide promising alternatives to transform food waste in high value products.

Besides that, anaerobic digestion and ethanol fermentation from food waste are effective biological processes to bioenergy that will help reducing of greenhouse gases. The application of food waste as a potential and sustainable feedstock for bioenergy on a commercial scale must be constructed by biorefinery concept.

However, the applications of food wastes in bioenergy must be improved on the efficiency increase of methane and/or hydrogen into biogas integrating different value-added bio-products and cost base by intensifying research and optimization studies with the ideal parameters of anaerobic processes.

REFERENCES

Adorno, M. A. T., Hirasawa, J. S. and Varesche, M. B. A. (2014). "Development and validation of two methods to quantify volatile acids (C2-C6) by GC/FID: headspace (automatic and manual) and liquid-liquid extraction (LLE)." *American Journal of Analytical Chemistry*, *5*, 406-414.

Arvanitoyannis, I. S. and Varzakas, T. H. (2008). "Vegetable waste treatment: Comparison and critical presentation of methodologies." *Critical Reviews in Food Science and Nutrition* 48:205-247. In: Kosseva, M.R. 2009. "Management and Processing of Food Wastes." *Advances in Food and Nutrition Research*, *58*, 57-136.

Banks, C. J., Chesshire, M., Heaven, S. and Arnold, R. (2011). "Anaerobic digestion of source-segregated domestic food waste: performance assessment by mass and energy balance." *Bioresource Technology*, *102*(2), 612-620.

Battimelli, A., Carrère, H. and Delgenès, J. P. (2009). S"aponification of fatty slaughterhouse wastes for enhancing anaerobic biodegradability." *Bioresource Technology*, *100*(15), 3695-3700.

Biernat, K. and Grzelak, P. L. (2015). "Biorefinery Systems as an Element of Sustainable." In *Biofuels – Status and Perspective*, 1 ed. Rijeka - Croatia IN Tech, *1*, 427-458.

Bouallagui, H., Ben Cheikh, R., Marouani, L. and Hamdi, M. (2003). "Mesophilic biogas production from fruit and vegetable waste in a tubular digester." *Bioresource Technology*, *86*(1), 85-89.

Contrera, R. C., Silva, K. C. C., Morita, D. M., Rodrigues, J. A. D., Zaiat, M. and Schalch, V. (2014). "First-order kinetics of landfill leachate treatment in a pilot-scale anaerobic sequence batch biofilm reactor." *Journal of Environmental Management*, *145*, 385-393.

Dague, R. R., Habben, C. E. and Pidaparti, S. R. (1992). "Initial studies on the anaerobic sequencing batch reactor." *Water Science and Technology*, *26*, 2429-2432.

Dung, T. N. B., Sen, B., Chen, C. C., Kumar, G. and Lin, C. Y. (2014). "Food Waste to Bioenergy via Anaerobic Processes." *Energy Procedia*, *61*, 307-312.

Ferraz, A. D. N., Koyama, M. H., Araújo, M. M. and Zaiat, M. (2016). "Thermophilic anaerobic digestion of raw sugarcane vinasse." *Renewable Energy*, *89*, 245-252.

Galanakis, C. M. (2012). "Recovery of high added-value components from food wastes: Conventional, emerging technologies and commercialized applications." *Trends in Food Science and Technology*, *26*(2), 68-87.

Gerard, M. H. (2006). *Wastewater Bacteria*. Wiley-Interscience, New Jersey. 255p.

Girotto, F., Alibardi, L. and Cossu, R. (2015). "Food waste generation and industrial uses: A review." *Waste Management*, *45*, 32-41.

Jain, S., Jain, S., Wolf, I. T., Lee, J. and Tong, Y. W. (2015). "A comprehensive review on operating parameters and different pretreatment methodologies for anaerobic digestion of municipal solid waste." *Renewable and Sustainable Energy Reviews*, *52*, 142-154.

Karmee, S. K. (2016). "Liquid biofuels from food waste: Current trends, prospect and limitation." *Renewable and Sustainable Energy Reviews*, *53*, 945-953.

Kim, D. H., Kim, S. H., Kim, K. Y. and Shin, H. S. (2010). "Experience of a pilot-scale hydrogen-producing anaerobic sequencing batch reactor (ASBR) treating food waste." *International Journal of Hydrogen Energy*, *35*(4), 1590-1594.

Kim, J. K., Han, G. H., Oh, B. R., Chun, Y. N., Eom, C. Y. and Kim, S. W. (2008). "Volumetric scale-up of a three stage fermentation system for food waste treatment." *Bioresource Technology*, *99*, 4394-4399.

Kim, J. K., Oh, B. R., Chun, Y. N. and Kim, S. W. (2006). "Effects of temperature and hydraulic retention time on anaerobic digestion of food waste." *Journal of Bioscience and Bioengineering*, *102*(4), 328-332.

Kim, S. H., Han, S. K. and Shin, H. S. (2004). "Feasibility of biohydrogen production by anaerobic co-digestion of food waste and sewage sludge." *International Journal of Hydrogen Energy*, *29*(15), 1607-1616.

Kiran, E. U., Trzcinski, A. P., Ng, W. J. and Lui, Y. (2014). "Bioconversion of food waste to energy: A review." *Fuel*, *134*, 389-399.

Kondusamy, D. and Kalamdhad, A. S. (2014). "Pre-treatment and anaerobic digestion of food waste for high rate methane production – A review." *Journal of Environmental Chemical Engineering*, *2*, 1821-1830.

Kosseva, M. R. (2009). "Processing of Food Wastes." *Advances in Food and Nutrition Research*, *58*, 57-136.

Kosseva, M. R. (2011). *Management and Processing of Food Wastes.* University College Dublin, Dublin, Ireland. Elsevier B.V.

Latif, M. A., Ahmad, A., Ghufran, R. and Wahid, Z. A. (2012). "Effect of temperature and organic loading rate on upflow anaerobic sludge blanket reactor and CH_4 production by treating liquidized food waste." *Environmental Progress and Sustainable Energy*, *31*(1), 114-121.

Lee, Z. K., Li, S. L., Kuo, P. C., Chen, I. C., Tien, Y. M., Huang, Y. J., Chuang, C. P., Wong, S. C. and Cheng, S. S. (2010). "Thermophilic bio-energy process study on hydrogen fermentation with vegetable kitchen waste." *International Journal of Hydrogen Energy*, *35*(24), 13458-13466.

Lissens, G., Vandevivere, P., De Baere, L., Biey, E. M. and Verstrae, W. (2001). "Solid waste digesters: process performance and practice for MSW digestion." *Water Science and Technology*, *44*(8), 91-102.

Maintinguer, S. I., Fernandes, B. S., Saavedra, N., Duarte, I., Adorno, M. A. T. and Varesche, M. B. A. (2008). "Fermentative Hydrogen Production by Microbial Consortium." *International Journal of Hydrogen Energy*, *33*, 4309-4317.

Maintinguer, S. I., Fernandes, B. S., Saavedra, N., Duarte, I., Adorno, M. A. T. and Varesche, M. B. A. (2011). "Fermentative hydrogen production with xylose by *Clostridium* and *Klebsiella* species in anaerobic batch reactors." *International Journal of Hydrogen Energy*, *36*, 13508-13517.

Maintinguer, S. I., Hatanaka, R. R. and De Oliveira, J. E. (2015). "Glycerol as a Raw Material for Hydrogen Production" In *Biofuels – Stats and Perspective*. 1 ed. Rijeka - *Croatia IN Tech*, *1*, 477-497.

Maintinguer, S. I., Sakamoto, I. K. T., Adorno, M. A. and Varesche, M. B. A. (2015). "Bacterial diversity from environmental sample applied to bio-hydrogen production." *International Journal of Hydrogen Energy*, *40*, 3180-3190.

Mata-Alvarez, J. M. (1987). "A dynamic simulation of a two-phase anaerobic digestion system for solid wastes." *Biotechnology and Bioengineering*, *30*, 844-851.

Mirabella, N., Castellani, V. and Sala, S. (2014). "Current options for the valorization of food manufacturing waste: a review." *Journal of Cleaner Production*, *65*, 28-41.

Murphy, J. D., McKeogh, E. and Kiely, G. (2004). "Technical/economic/ environmental analysis of biogas utilization." *Applied Energy*, *77*(4), 407-427.

Okada, D. Y., Delforno, T. P., Esteves, A. S., Sakamoto, I. K., Duarte, I. C. S. and Varesche, M. B. A. (2013). "Optimization of linear alkylbenzene sulfonate (LAS) degradation in UASB reactors by varying bioavailability of LAS, hydraulic retention time and specific organic load rate." *Bioresource Technology*, *128*, 125-133.

Oliveira, R. P., Ratusznei, S. M., Rodrigues, J. A. D., Zaiat, M. and Foresti, E. (2010). "Interaction effects of organic load and cycle time in an AsBr applied to a personal care industry wastewater treatment." *Journal of Environmental Management*, *91*, 2499-2504.

Park, Y. J., Hong, F., Cheon, J. H., Hidaka, T. and Tsuno, H. (2008). "Comparison of thermophilic anaerobic digestion characteristics between single-phase and two-phase systems for kitchen garbage treatment." *Journal of Bioscience and Bioengineering*, *105*(1), 48-54.

Pfaltzgraff, L. A. and Clark, J. H. (2014). "1 - Green chemistry, biorefineries and second generation strategies for re-use of waste: an overview." *Advances in Biorefineries*, 3-33.

Pham, T. P. T., Kaushik, R., Parshetti, G. K., Mahmood, R. and Balasubramanian, R. (2015). "Food waste-to-energy conversion technologies: Current status and future directions." *Waste Management*, *38*, 399-408.

Seghezzo, L., Zeeman, G., Van Lier, J. B., Hamelers, H. V. M. and Lettinga, G. (1998). "A review: The anaerobic treatment of sewage in UASB and EGSB reactors." *Bioresource Technology*, *65*(3), 175-190.

Sen, B., Aravind, J., Kanmani, P. and Chyi-How, L. (2016). "State of the art and future concept of foodwaste fermentation to bioenergy." *Renewable and Sustainable Energy Reviews*, *53*, 547-557.

Shin, H. S., Youn, J. S. and Kim, S. H. (2004). "Hydrogen production from food waste in anaerobic mesophilic and thermophilic acidogenesis." *International Journal of Hydrogen Energy*, *29*(13), 1355-1363.

Sitorus, B. and Sukandarb Panjaitanc, S. D. (2013). "Biogas recovery from anaerobic digestion process of mixed fruit-vegetable wastes." *Energy Procedia*, *32*, 176-182.

Speece, R. E. (1996). *Anaerobic Biotechnology for Industrial Wastewaters.* Archae Press, Nashville, TE, USA, 394-400.

Youn, J. and Shin, H. (2005). "Comparative performance between temperature-phased and conventional mesophilic two-phased processes in terms of anaerobically produced bioenergy from food waste." *Waste Manage Resources*, *23*(1), 32-38.

Zhang, C., Su, H., Baeyens, J. and Tan, T. (2014). "Reviewing the anaerobic digestion of food waste for biogas production." *Renewable and Sustainable Energy Reviews*, *38*, 383-392.

In: Food Waste
Editor: Garrett Leonard Riley

ISBN: 978-1-63485-025-4

Chapter 5

ANAEROBIC DIGESTION: A PROMISING SOLUTION FOR FOOD WASTE MANAGEMENT IN URBAN AREAS IN VIETNAM

Hoa Huu Nguyen*
Visiting Researcher at the University of Southampton, UK (UoS)
National University of Civil Engineering, Vietnam
(NUCE)

ABSTRACT

In Vietnam, the generation of Municipal Solid Waste (MSW) is growing rapidly, in parallel with urbanisation. The composition of MSW is mainly food waste which is a highly biodegradable material and can be digested with Anaerobic Digestion (AD) to produce biogas. This can then be used to generate heat, electricity, and biofuels. However, in Vietnam food waste is mostly treated with other components in the MSW stream by landfilling. The disadvantages of this treatment can be clearly seen: direct landfilling of MSW has been known to create lasting detrimental impacts to the environment (emissions to the atmosphere, hydrosphere, risk in landfill stability, and scarcity of land). In comparison AD is a better method to treat biogenic wastes from an ecological and economic point of view. It is also receiving increased attention as a promising option for energy recovery to help mitigate against energy shortages.

* Corresponding Author's Email: hhn1g10@soton.ac.uk.

However, applying AD technology has been neglected by the Vietnam government when formulating national strategies to deal with MSW problems due to a lack of information, data, and experience. This manuscript reviews conventional methods currently applied in Vietnam for treatment of food waste, and then proposes a promising solution using anaerobic digestion technology by providing an overview of the anaerobic digestion development worldwide, estimation of energy potential from food waste in the municipal solid waste stream of urban areas in Vietnam through AD in forms of heat, electricity, and upgraded biogas.

The data and information from the manuscript can be employed to help Vietnam government and industrial decision makers to establish more effective and sustainable MSW management strategies. Also, it might be used as a reference information for other tropical countries like Vietnam.

Keywords: anaerobic digestion, food waste, Aspen Plus, energy, biogas, Vietnam

1. Introduction

1.1. Population Growth and Increased Urbanization in Vietnam

Vietnam is a tropical country with a total area of approximately 331,210 km^2 located on the eastern Indochina Peninsula. Urbanisation in Vietnam has been increasing rapidly along with the country's economic growth. In 2000, the number of cities and towns in Vietnam was 649 but this has increased to 715 in 2005 and 774 in 2014. The growth in the number of people moving from rural to urban areas is the main driver in the expansion of Vietnam's urban population. As of 2009, the urban population was 25.59 million, accounting for 29.74% of the total population; this expanded to 26.22 million (30.17%) a year later.

According to MONRE (2011), it has been estimated that in 2025 the urban population in Vietnam will double to 52 million, accounting for 50% of Vietnam's total population. This rapid urbanisation has put pressure on the government in dealing with environmental problems, including solid waste management in cities; and this is exacerbated by the limited space for the treatment of waste by traditional methods. For Vietnam to develop sustainably more effort is required to solve this challenge.

1.2. QUANTITY AND COMPOSITION OF MSW IN VIETNAM

In Vietnam, the generation of MSW is growing rapidly in parallel with urbanisation. According to recent reports, urban areas currently contain 30% of the country's population but produce about 42-46% of total solid wastes. The waste is mainly generated from households, buildings, commercial activities and commercial enterprises such as offices, hotels, retail, institutions, and from municipal services such as street cleaning, etc.

Table 1. Composition (%) of MSW from some of the main cities in Vietnam 2009 – 2010

Components	Hanoi (Nam Son)	Hanoi (Xuan Son)	Haiphong (Trang Cat)	Haiphong (Dinh Vu)	Hue (Thuy Phuong)	Danang (Hoa Khanh)	HCM (Da Phuoc)	HCM (Phuoc Hiep)	Bacninh (Ho)	Average
Food waste *	53.81	60.79	55.18	57.56	77.1	68.47	64.50	62.83	56.9	61.9
Paper	6.53	5.38	4.54	5.42	1.92	5.07	8.17	6.05	3.73	5.20
Textile	5.82	1.76	4.57	5.12	2.89	1.55	3.88	2.09	1.07	3.19
Wood	2.51	6.63	4.93	3.70	0.59	2.79	4.59	4.18	-	3.74
Plastic	13.57	8.35	14.34	11.28	12.47	11.36	12.42	15.96	9.65	12.15
Leather and Rubber	0.15	0.22	1.05	1.90	0.28	0.23	0.44	0.93	0.20	0.60
Metal	0.87	0.25	0.47	0.25	0.40	1.45	0.36	0.59	-	0.58
Glass	1.87	5.07	1.69	1.35	0.39	0.14	0.40	0.86	0.58	1.37
porcelain	0.39	1.26	1.27	0.44	0.79	0.79	0.24	1.27	-	0.81
Soil and Sand	6.29	5.44	3.08	2.96	1.70	6.75	1.39	2.28	27.85	6.42
cinder	3.10	2.34	5.70	6.06	-	0.00	0.44	0.39	-	2.58
hazardous waste	0.17	0.82	0.05	0.05	-	0.02	0.12	0.05	0.07	0.17
Sludge	4.34	1.63	2.29	2.75	1.46	1.35	2.92	1.89	-	2.33
Others	0.58	0.05	1.46	1.14	-	0.03	0.14	0.04	-	0.49
Total	100	100	100	100	100	100	100	100	100	

Note: * food waste can be considered as the organic fraction of MSW when it does not contain irrecoverable paper residues residues (Zhu et al. 2008, Hartmann and Ahring 2005). The data have been adapted from JICA (2011) and MONRE (2011).

MSW in Vietnams urban areas is mainly composed of food waste, paper, plastic, wood, metal and glass, with some hazardous household waste such as electric light bulbs, batteries, etc. (MONRE 2011). Composition of MSW from some of the main cities in Vietnam between 2009 and 2010 is presented in Table 1.

As can be seen from Table 1, food waste accounts for a very large proportion of the MSW stream, ranging between 54-77% dependent on the city.

The proportion of food waste can be estimated based on the municipal waste generation rate, population growth in urban areas and the organic fraction. Table 2 shows estimates for the predicted generation of food waste in cities for 2015, 2020 and 2025.

Table 2. Estimation of MSW and food waste generation in Vietnam for the 2015, 2020 and 2025

Contents	2015	2020	2025
Urban population (million)[a]	35	44	52
% of Vietnam's population	38	45	50
Municipal waste generation rate (kg $cap^{-1}day^{-1}$)[b]	1.2	1.4	1.6
Total MSW per day (tonnes day^{-1})	42,000	61,600	83,200
Rate of MSW collected (% of total MSW)	85[c]	90[c]	100[d]
Total MSW collected per day (tonnes day^{-1})	35,700	55,440	83,200
Total food waste collected per day (tonnes day^{-1})[e]	21,420	33,264	49,920

[a, d] data taken from the GWP (2009).
[b] according to the MONRE (2011).
[c] according to the GWP (2011).
[e] assumed that food waste fraction in MSW stream is 60%.

1.3. Current Treatment Methods for Solid Wastes in Urban Areas in Vietnam and the Problems

Currently, there are four main methods used in Vietnam to treat MSW as shown in Figure 1.

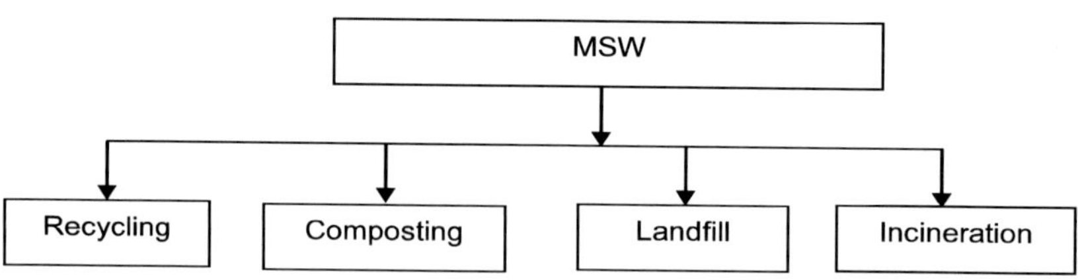

Figure 1. Popular methods currently used for MSW treatment in Vietnam (MONRE 2011).

Out of these options landfilling is the most common treatment method which is applied widely in cities across Vietnam to deal with almost all types of MSW. It is considered the simplest, and in many cases the cheapest, method of disposal. During the stabilisation phase of the landfill process the solid waste produces leachate but also landfill gas, which can be used for heat and electricity generation. This process has many significant economic and environmental problems, however, including: limited space for landfill sites; cost of waste burial; cost of transporting the waste; risk of contamination of groundwater with pollutants; and the emission of greenhouse gases to the atmosphere (Byrne, 1997; Trzcinski and Stuckey, 2009). According to data published in October 2015, only 17 of the 91 landfill sites in Vietnam are sanitary. Furthermore, due to space limitations landfill sites will have to be located further from the cities, considerably increasing transportation costs.

The second most common method of waste disposal is composting. This involves the decomposition of organic matter under controlled conditions to produce a nutrient rich product that can be used as a soil conditioner or fertilizer (Cheng and Hu 2010). The high organic matter content of the MSW in urban areas in Vietnams provinces have a great potential for composting. However, the method is not widely applied in Vietnam due to a lack of knowledge and experience as well as various drawbacks. The main downside of this technology is the cost for site preparation and equipment. It requires more space than other waste management technologies and preparation of the material to reduce the particulate size, as well as some degree of waste separation/processing. Other problems with composting are due to the relatively lengthy treatment period required to produce the final compost product, and environmental issues such as the generation of odour or contamination of heavy metals and pathogens if not managed properly (Nguyen and Le 2011, Cheng and Hu 2010).

Incineration is not applied widely in Vietnam. In most cases, incineration is a pre-treatment step for the hygienically safe destruction of toxic medical waste arising from hospitals before it can be disposed to landfill. However, due to the poor quality of incinerators and the lack of adequate technologies to analyse dioxin concentrations, the health effects of dioxin and furan emissions from incinerators; especially during start up and shut down are still considered as a problem. Existing small-scale incinerators, with capacities between 5-20 tons per day are a large problem, as they cannot meet the national standards for dioxins and furans emission (Nguyen and Le 2011).

Open burning at improper landfill sites is an illegal practice but still used in the dry season to reduce the volume of waste and increase the capacity of disposal sites. It is estimated that annually about 40-50% of total waste at these landfills is burned. This would have a detrimental impact on the environment and human health.

There is no simple solution to the waste disposal issue in Vietnam, but it requires alternative methods to shift the current paradigm.

2. Anaerobic Digestion Method for MSW Treatment and Energy Recycling

2.1. Background of the Anaerobic Digestion Technology

Anaerobic digestion occurs in Environments depleted of oxygen, and involves the breakdown of organic matter into biogas and other trace gases, as well as a residual effluent or digestate. There are four main steps in the anaerobic digestion process: hydrolysis, acidogenesis, acetogenesis and methanogenesis. In the first process, hydrolysis, complex materials are converted into less complex and soluble compounds. In the acidogenesis phase, volatile fatty acids (VFAs) are generated alongside alcohols, lactic acid, CO_2, H_2, NH_3, H_2S and new cellular material. Acetogenesis is the third step with the production of acetate and molecular hydrogen via the anaerobic oxidation of higher fatty acids and the conversion of propionate, butyrate and valerate to acetate and hydrogen (acetogenesis process). Methanogenesis, the final stage, involves the production of methane from the materials produced in previous stages. A schematic outline of the digestion process is presented in Figure 2.

Hydrolysis: during the hydrolysis process, proteins are hydrolysed to amino acids, polysaccharides to simple sugars and lipids to long chain fatty acids by the action of exo-enzymes produced by hydrolytic bacteria. These products are monomeric and dimeric compounds which are readily accessible for acidogenic bacteria in the next stage.

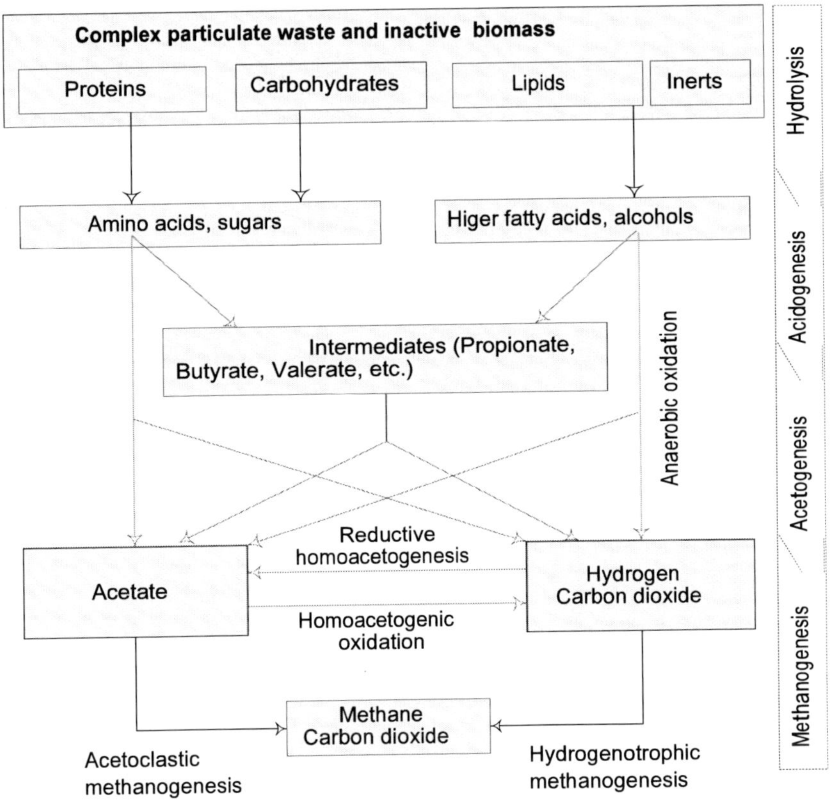

Figure 2. A schematic pathway of anaerobic conversion from biomass to methane. Adapted from Batstone et al. (2002); Lier, Mahmoud, and Zeeman (2008), Demirel and Scherer (2008).

Acidogenesis: this is generally considered to be the fastest step in the anaerobic digestion process. Growth rates of acidogenic bacteria are ten to twenty-fold higher than those of methanogens, and bacterial yields and conversion rates are five-fold higher. Products from this stage cannot be used directly by the methanogens and must be degraded further in a subsequent process, namely, acetogenesis (Bjornsson 2000).

Acetogenesis: in the acetogenesis stage, which is carried out mainly by obligate hydrogen producing acetogens (OHPA), VFAs and LCFAs are syntrophically oxidised to produce acetate, carbon dioxide and hydrogen. These are the only substrates that can be metabolised efficiently by the methanogens in the final stage of anaerobic digestion (Anderson, Sallisp, and Uyanik 2003). New cell materials also are created in the acetogenesis process (Lier, Mahmoud, and Zeeman 2008).

Table 3. Benefits of AD (gathered from Stuckey (1986) Fulford (1988), Maier, Pepper, and Gerba (2000), Spellman (2007), Deublein and Steinhauser (2008), Wall, Harwood, and Demain (2008), Sivanagaraju (2010) and Conly (2011)

Energy benefits	• Produces a renewable fuel that is flexible and can be used to generate heat and power. • Biogas can be used on site, or upgraded to biomethane for use as a vehicle fuel, or transported to where energy is demanded e.g., by injection into the gas grid. • Removing the wet fraction from waste improves the calorific value of the residual waste stream.
Environmental benefits	• Reduction of greenhouse gas emissions by capture and use of methane that might otherwise leak into the atmosphere and increase the greenhouse effect. • Can reduce emissions of nitrous oxide (a strong greenhouse gas) compared to composting or landfill. • Can contribute to improved nutrient management creating a closed cycle of nutrients (N, P, K). • Plant fertiliser quality is improved compared to raw agricultural waste • Good pathogen removal depending on temperature.
Waste treatment benefits	• Allows wastes to be treated locally and at small scale, adhering to the proximity principle. • Potential for co-digestion with other organic waste streams (industrial wastes e.g., food processing waste and agricultural wastes e.g., manure)
Economic benefits	• No supply costs in the case of waste products utilisation • Substantial reduction of the disposal costs of organic wastes, even including meaningful re-use (e.g., as fertilisers), because the quantity of biomass decreases so significantly • Digested residues and fibre can potentially be used or sold.

Methanogenesis: the methanogenic step is the final stage of the conversion process of organic matter, with the two important products being methane and carbon dioxide. There are two routes of methane formation. In the first route, acetate is cleaved to form carbon dioxide and methane (acetoclastic methanogenesis). In the second route, acetate is oxidised to H_2 and CO_2, then CO_2 is reduced with H_2 to yield CH_4 and H_2O (the syntrophic acetate oxidation pathway).

In the anaerobic digestion system, a healthy population of the relevant groups of microorganisms is a very important factor for effective degradation of organic materials. A large number of studies have been carried out on factors affecting the anaerobic digestion process such as substrate, nutrients and trace elements, temperature, pH and alkalinity, proper mixing, toxicity and inhibition. In tropical countries such as Vietnam, it is claimed that the average annual temperature is preferable for the anaerobic digestion process.

The benefits of anaerobic digestion can be categorised under four major aspects: energy, environmental, waste treatment, and economic, as summarised in Table 3.

2.2. Development of Anaerobic Digestion Worldwide and in Vietnam

The AD process is widely applied as an effective option for renewable energy because under controlled conditions the biogas produced (consisting mainly of methane and carbon dioxide) can be used for energy production, helping to replace fossil fuels. The nitrogen content and minerals in the digestate make it potentially useful for agricultural application as a soil conditioner or bio-fertiliser with associated economic, energy and carbon gains from offsetting the requirement for artificial fertilisers.

According to He (2010), there is historical evidence that the anaerobic digestion process was used in China and India about 2000-3000 years ago. The history of scientific interest in anaerobic digestion perhaps dates back to 1776 when Volta made an estimate of the amount of flammable gas produced by decaying organic materials (Fujishima, Miyahara, and Noike 2000). This was later assessed and corrected by Dalton, Henry and Davy between 1801 to 1810 (Gunnerson and Stuckey 1986, Appels et al. 2011). The past few decades have seen increasing interest in both research and real-life applications of AD. From the 1990s to the present, research on anaerobic digestion has developed rapidly, with one driving force being the desire of many governments to address both rising energy demands and the need for renewable energy sources. Moreover, from the ecological and economic points of view AD is a better option for management of biodegradable wastes when compared to conventional methods e.g., landfilling and composting (Edelmann, Schleiss, and Joss 2000). The European Commission (2010) stated that proper handling of biowastes, such as the use of AD technology, is often the best and most cost-effective means of addressing a range of other problems from nutrient management to greenhouse gas emissions.

In developing countries AD technology also flourishes as a waste-to-energy solution, but on a smaller scale and in a decentralized manner. For example in China, in 2002, there were about 11 million digesters, this had almost doubled to more than 19 million by 2006 (Deublein and Steinhauser 2011). In India, the total number of family size biogas plants installed in 2005 was under 4 million, compared to 12 million in 2010 (Rao et al. 2010).

In Vietnam, biogas production was introduced over 15 years ago. By the end of 2006, more than 18,000 domestic biogas plants had been installed in 10 provinces in Vietnam with support from the government of the Netherlands. This investment, however, was only at the household or household-group levels for the treatment of agricultural wastes, manure, etc. (Abbasi, Tauseef, and Abbasi 2012). When it comes to MSW, AD has been neglected by the government. The main reason for this is a lack of information, data and experience, as well as economic factors. To date, AD does not yet make a significant contribution to resolving Vietnam's urban waste issues. The National Environment Report of Vietnam (2011) indicated that strategies from now to 2025 will focus on methods to recover energy and materials from MSW in cities. Recently, the Vietnam Government again emphasised that it is necessary to develop waste management systems in which solid wastes are classified at source, collected, reused, renewed and treated with progressive technologies to boost technological innovation in waste-to-energy processes (GWP 2012). Thus, waste-to-energy technologies could not only help to provide a solution to this problem but also meet the national energy consumption policy for sustainable development (MONRE 2011).

2.3. Food Waste as a High Potential Energy Source for AD

Food waste is a typical type of organic matter containing high potential for energy production through anaerobic degradation. Its characteristics can be noticeably variable depending upon their sources. Some characteristics of food wastes that have been reported in the literature are presented in Table 4, showing moisture content of 70–90%, volatile solids to total solids ratio (VS/TS) of 85–95%, and carbon to nitrogen ratio (C/N) of 9–36.4.

Table 4. Characteristics of food wastes listed in literature

Source	Characteristics			Countries	Sources of reference
	Moisture content (%)	VS/TS (%)	C/N ratio		
A dining hall	79.5	95	14.7	Korea	Han and Shin (2004)
University's cafeteria	80.03	93.55	NA*	Korea	(Kwon and Lee 2004)
A dining hall	NA	94	18.3	Korea	(Shin, Youn, and Kim 2004)
A dining hall	84.1	95.6	NA	Korea	(Kim, Han, and Shin 2004)
University's cafeteria	87.6	89.3	9.2	Korea	Kim et al. (2008)
Segregated domestic food waste	77	92	14	England	Charles J Banks, Michael Chesshire, and Stringfellow (2008)
Separated from MSW of City of San Francisco	69.1	85.43	14.8	USA	Zhang et al. (2007b)
University restaurant	81.9	94.47	13.2	Korea	Zhang, Lee, and Jahng (2011)
Emanating from fruit and vegetable markets, household and juices centres	85	88.5	36.36	India	Rao and Singh (2004)

* NA – Not available.

Studies have shown that the methane yield from anaerobic digestion is in the range of 350 to 435 mL g^{-1} VS_{added} depending upon operational conditions, reactor types, and composition of the input food waste (Banks et al. 2011, Cho, Park, and Chang 1995, Zhang et al. 2007a, Demirel et al. 2010). Finnveden et al. (2005) analysed data obtained from LCA scenarios and concluded that with food waste substrate, in general, anaerobic digestion is preferable to composting and landfilling regarding energy use and emissions of greenhouse gases. Because of the benefits in terms of energy saving, waste management and environmental aspects, biogas production from food waste together with other renewable organic sources i.e., agricultural waste has been suggested as a means of meeting one-third of renewable energy demand in transport in the EU by 2020 (COM 2010).

It has been claimed that although AD can potentially offer many benefits as shown, its application to certain types of organic waste can be difficult to implement because it is a complex system of biochemical and physical processes. However, recent studies on the anaerobic digestion of organic wastes have indicated that using anaerobic digestion to deal with food waste is an effective and promising option from both environmental and energetic points of view (Banks and Zhang 2010, Zhang et al. 2007a, Banks et al. 2011, Banks et al. 2012, Nguyen, Heaven, and Banks 2014, Nguyen 2014).

3. Estimation of Energy Potential from MSW Stream in Urban Areas in Vietnam

To date, sources of information about the achievable energy from food waste created in cities in Vietnam using AD technology is limited (Nguyen, Heaven, and Banks 2014). The aim of this section is to provide estimates of the energy potential from food waste in the MSW from urban areas in Vietnam through AD in forms of heat, electricity and upgraded biogas using an energy model developed at the University of Southampton, UK. In order to achieve this the steps detailed below were carried out:

-
- Investigate, collect data from published National Environment Reports.
- Estimate food waste generation based estimation of population in the next milestones and strategies on food waste management of urban area system in Vietnam up to 2025.
- Use the developed model to estimate potential energy from waste stream in accordance with urbanisation strategies.

Results from estimations with different scenarios can be used as a source of data for evaluation of the AD method as a treatment of food waste in MSW streams from the energy recovery point of view. Furthermore, this could help the Vietnamese government and industry decision-makers to establish more effective and sustainable MSW management strategies.

3.1. Anaerobic Digestion Energy Balance Model

A mass and energy balance model of the anaerobic digestion system based on stoichiometric approach was built in Aspen Plus flowsheeting software (AspenTech 2011). The energy model used in this work incorporates an updated investigation on food waste aspects conducted by researchers from the University of Southampton, UK and partners in Austria, Finland and India. Details of components in system such as mixing, heating, biogas upgrading, etc. can be found in (Nguyen 2014).

Figure 3 shows the main elements of an AD plant which have been taken into account in the model.

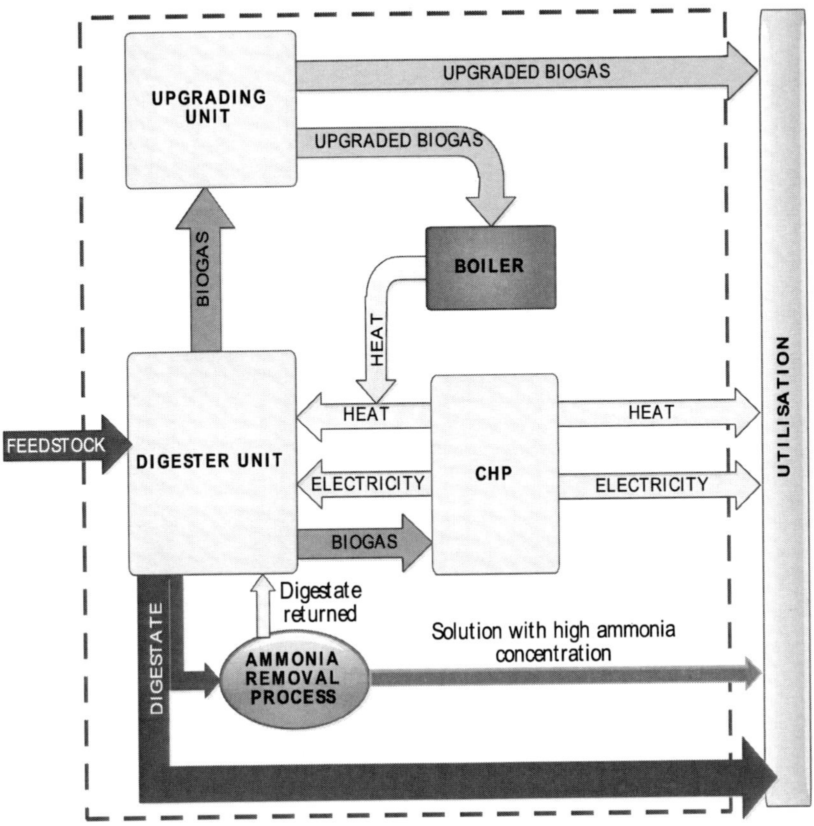

Figure 3. System boundaries (inside the dashed line - - - - - - -).

There are four main components in this model, including digester unit, CHP unit, upgrading unit and boiler unit. In order to make this model simpler, basic assumptions have been made:

- The amount of biogas produced can be predicted using Buswell's Equation (Buswell and Neave 1930) which is still in use worldwide.

The empirical formula of food waste was estimated based on its ultimate analysis and can be can be expressed as: $C_{3.963}H_{7.608}O_{2.248}N_{0.248}S_{0.005}$ (Nguyen 2014).

- The digesters work at optimum conditions (adequate temperature, pH, mixing, etc.),
- The main impurities are H_2S and NH_3, other trace gases are minor and can be neglected.

3.2. Scenarios, Results and Discussion

3.2.1. Setting Scenarios

The estimated amounts of food waste generated in 2015, 2020 and 2025 as shown in Table 2 were used in this model.

Table 5. AD plant operational parameters

Parameters	Unit	Values
Digester temperature	^oC	35
Pasteuriser temperature	^oC	70
Ambient temperature	^oC	22/27
Total food waste (wet)	$x10^3$ kg day^{-1}	21420/22260/49920
Time in pasteuriser	hour	1
Loading rate	kg VS m^{-3} day^{-1}	3
TS in food waste	%	27.8
VS in food waste	%	25
Overall heat transfer coefficient of digesters surroundings		
Wall	W m^{-2} K	0.275
Floor	W m^{-2} K	0.823
Roof	W m^{-2} K	0.931

The seasonal average temperature in Vietnam was taken as 22°C in winter and 27°C in summer (Usa, 2007).

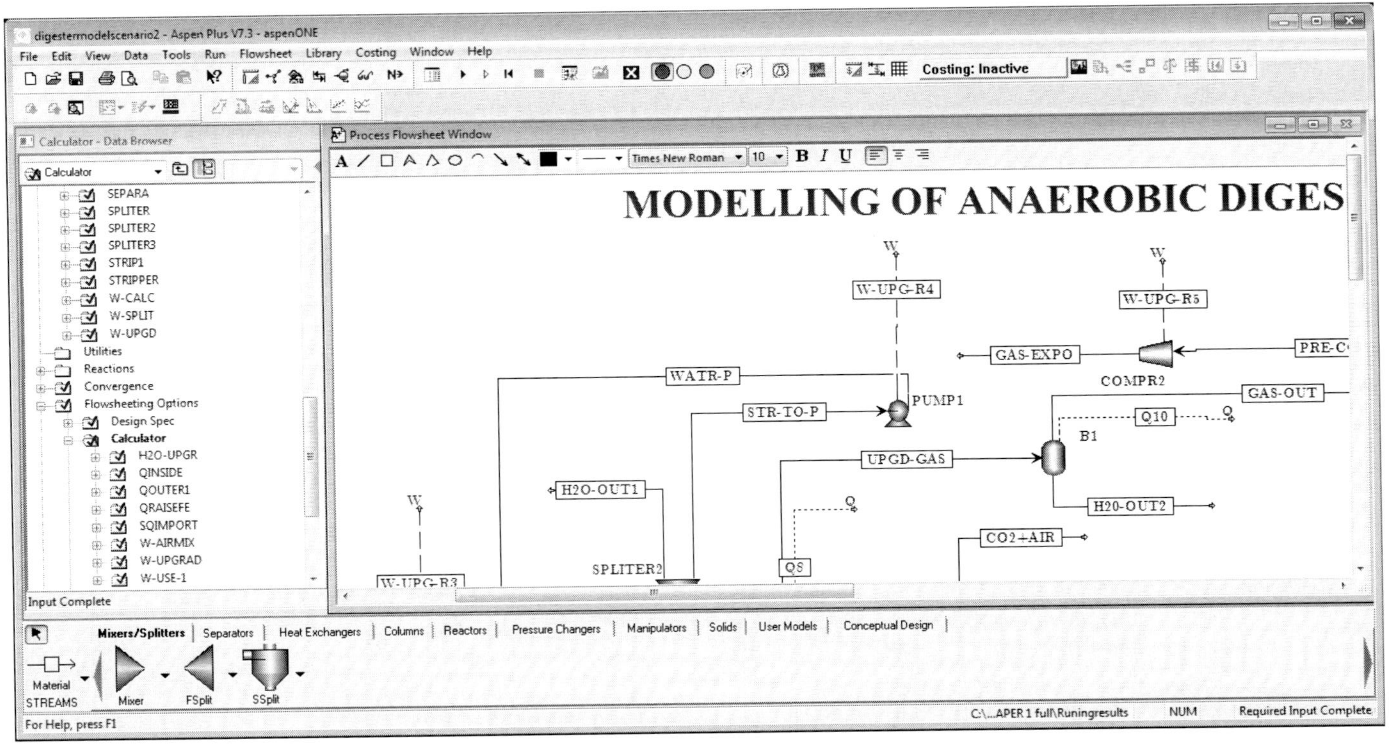

Figure 4. Model screenshot.

As already mentioned, biogas can be utilised in the forms of heat, electricity, vehicle fuel, natural gas, fuel cells, etc. Among these, electricity and vehicle fuel are currently the most suitable for use in Vietnam, and therefore these two were used in the scenarios to estimate the possible energy from food waste. These scenarios consider parameters which directly affect the overall energy consumption in operation of the AD plant, such as: ambient temperature, total input food waste, heat loss, etc. The required digester volume is also increased by 10% for gas storage. Other assumptions made in the model are summarised in Table 5.

Scenario 1 was chosen to obtain the maximum possible usable energy in the form of heat and electricity, by sending all of the biogas generated from the digesters to the CHP unit. The surplus heat and electricity after internal uses (for pumps, mixers, compressors, etc.) is then sold for commercial purposes.

3.2.2. Results and Discussion

The amount of biogas and digestate derived from the AD of food waste for different years are shown in Table 6.

Table 6. Biogas generated, gas lost and digestate from food waste in years

Contents	Unit	2015	2020	2025
Food waste	tonnes day^{-1}	21420	22260	49920
Gas generated	tonnes day^{-1}	5858	6088	13654
Biogas available	tonnes day^{-1}	4221	4397	9853
Digestate	tonnes day^{-1}	15561	16171	36265

As can be seen from Table 6, the biogas generated in 2025 is about 9850 tonnes day^{-1} which is two times greater than in 2015.

Results from running scenario 1 (Table 7) show that energy in the form of heat produced from the CHP unit is about two times greater than the electricity generated. This reflects the fact that the efficiency of heat from CHP units is about 65% whereas in electric power it is around 35-43% (Verougstraete et al., 1985; Weiland, 2010; Spellman, 2013).

From these figures the electricity generated which can be sent to the grid could contribute between 2.4% to 4.1% of the total electricity demand in Vietnam in 2015 and 2025, respectively. The heat produced can also be used for heating/cooling in buildings (Deng et al., 2011; Choudhury et al., 2013) or other industrial purposes in urban areas.

Table 7. Scenario 1: Energy potential in the form of heat and electricity

Contents	Unit	2015		2020		2025	
Ambient temp.		Winter	Summer	Winter	Summer	Winter	Summer
Heat and electricity for off-site use							
Heat	GWh	12.6	13.0	13.1	13.5	28.2	29.0
Electricity	GWh	6.85	6.85	7.10	7.10	14.8	14.8
Annual electricity demands of VN*	GWh		241		301		360

Note: *adapted from World Bank (2012).

In the second scenario (Table 8), upgraded biogas with a methane content of over 97% (clean gas) can be used as fuel for trucks, buses, etc. According to Vietnam's national energy development strategy (2007), it is estimated that the fuel requirement for transportation in 2020 will be 23 million tonnes, increasing to around 30 million tonnes in 2025. Assuming 1 kg fuel (diesel) is equal to 11.5 kWh (Denny, 2013) then upgraded biogas can replace about 2.2% and 4.75% of anticipated daily fuel use for transportation in 2020 and 2025, respectively.

Table 8. Scenario 2: Energy potential in the form of upgraded biogas

Contents	Unit	2015		2020		2025	
Ambient temp.		Winter	Summer	Winter	Summer	Winter	Summer
Heat and electricity for off-site use							
Heat	GWh	3.06	3.74	3.23	3.90	7.50	9.09
Energy potential of upgraded biogas	GWh	16.03	16.11	16.85	16.74	37.33	37.36

Because of the high average ambient temperature, the heat requirements for internal uses such as for heating the digester, pasteurising etc. are small compared to those in cooler climates (Smyth, Murphy, and O'Brien 2009, Pertl, Mostbauer, and Obersteiner 2010, Deublein and Steinhauser 2008). Therefore, use of the surplus biogas for on-site electricity generation in a CHP plant will produce a large amount of surplus heat. This is potentially available for export but it is notoriously difficult to find economic uses for it, especially in warm climates where there is little or no demand for domestic heating. Waste heat can, however, be used for cooling and industrial purposes and work on development of these areas is important to ensure effective use of the renewable energy.

4. Proposal of MSW Segregation, Flow and Treatment for Urban Areas in Vietnam Including Anaerobic Digestion

Results from running scenarios in an energy balance model show that if the food waste in the MSW stream from cities in Vietnam could be separated, it could be a significant source of energy in the form of heat, electricity or biofuel. This can potentially be achieved in the future through encouraging changes in people's behaviour and enforcement of environmental laws. The total surplus exportable energy generated in Vietnam from biogas plants (working at standard conditions) each day in any energy form (heat, electricity or purified gas)is estimated to be approximately 19 GWh, 20 GWh and 45 GWh in 2015, 2020 and 2025, respectively. This work also reveals that when food waste is separated from the MSW stream and sent to AD plants, it could contribute between 2.4% to 4.1% of the electricity demand of Vietnam, with about double this energy also in the form of heat. Alternatively, upgrading this biogas could contribute approximately 2.2% to 4.7% of fuel consumption for transportation. This suggests AD is a promising method to treat MSW in cities, especially when considering the problematic aspects of other waste disposal methods such as: landfilling, composting and incineration.

In the future, it is suggested that the anaerobic digestion should be included in Vietnams strategies for MSW management and it should play an important role in dealing with the food waste fraction in the MSW stream. Based on technologies being used in Vietnam and the potential of anaerobic digestion, a systematic scheme for MSW management in urban areas of Vietnam integrated with AD method can be proposed as shown in Figure 5.

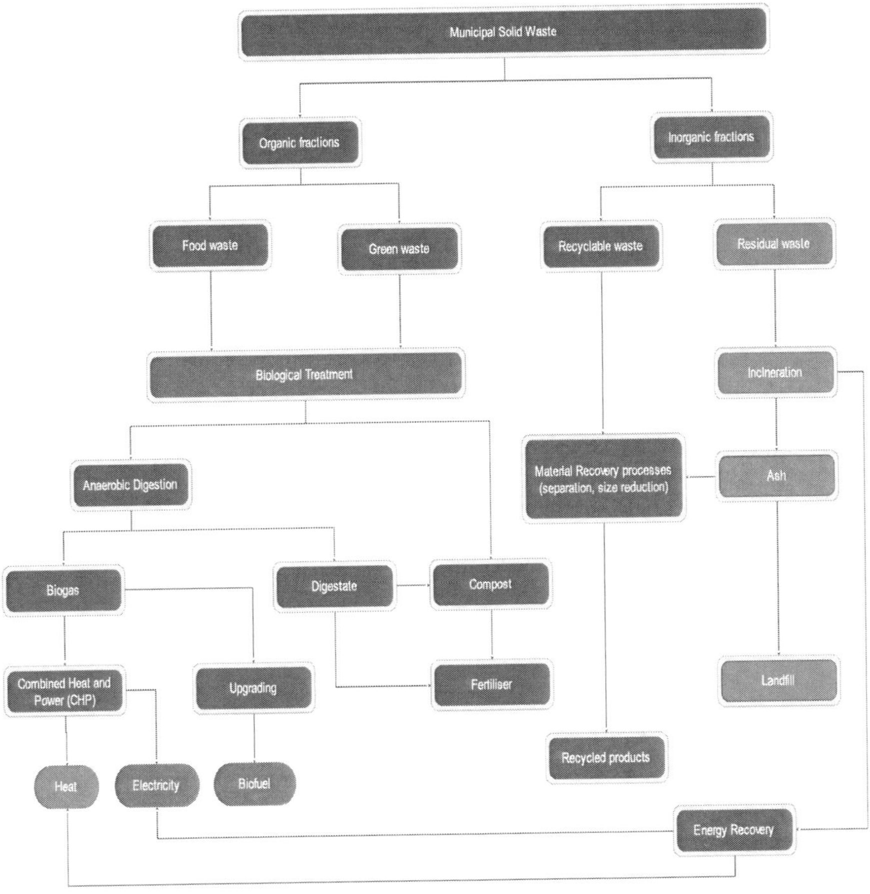

Figure 5. Proposal of MSW management in urban areas of Vietnam integrated with AD.

Conclusion

Due to the strong economic growth and urbanisation in recent years, Vietnam faces many environmental challenges. In particular solid waste management in cities has been promoted as the big issue. Solid waste generation in Vietnam is increasing dramatically, mainly generated from households, commercial activities such as wastes from offices, hotels, supermarkets, shops, institutions, and from municipal services such as street cleaning, etc. The main component of MSW is food waste which has a very high potential for bioenergy production.

If the food waste component in the MSW stream of cities in Vietnam could be separated, then AD offers a source of energy generation with many economic benefits. This could provide a contribution towards dealing with the dramatic increases in costs associated with energy supply, waste disposal, space for landfilling and the increasing public concerns with environmental issues.

REFERENCES

Abbasi, Tasneem, SM Tauseef, and SA Abbasi. 2012. "Biogas Capture from Solid Waste." In *Biogas Energy*, 105-143. Springer.

Anderson, GK., P. Sallisp, and S. Uyanik. 2003. "Anaerobic treatment processes." In *Handbook of Water and Wastewater Microbiology*, 391-426. London; San Diego: Academic Press.

Appels, L., J. Lauwers, J. Degreve, L. Helsen, B. Lievens, K. Willems, J. Van Impe, and R. Dewil. 2011. "Anaerobic digestion in global bio-energy production: Potential and research challenges." *Renewable and Sustainable Energy Reviews* 15 (9):4295-4301.

AspenTech. 2011. "Aspen Plus V7.3; Aspen Technology, Inc."

Banks, CJ., M. Chesshire, S. Heaven, and R. Arnold. 2011. "Anaerobic digestion of source-segregated domestic food waste: Performance assessment by mass and energy balance." *Bioresource Technology* 102 (2):612-620. doi: 10.1016/j.biortech.2010.08.005.

Banks, CJ., and Y. Zhang. 2010. Technical Report: Optimising inputs and outputs from anaerobic digestion processes. Defra Project Code WR0212. 2010.

Banks, CJ., Y. Zhang, Y. Jiang, and S. Heaven. 2012. "Trace element requirements for stable food waste digestion at elevated ammonia concentrations." *Bioresource Technology* 104:127-135. doi: DOI 10.1016/j.biortech.2011.10.068.

Batstone, DJ., J. Keller, I. Angelidaki, S. Kalyuzhny, S. Pavlostathis, A. Rozzi, W. Sanders, H. Siegrist, and V. Vavilin. 2002. *Anaerobic digestion model no. 1 (ADM1)*: IWA publishing.

Bjornsson, Lovisa. 2000. "Intensification of the biogas process by improved process monitoring and biomass retention." [Univ.-bibl.]; http://worldcat.org.

Buswell, AM., and SL. Neave. 1930. *Laboratory studies of sludge digestion, Illinois Water survey Bulletin*. Springfield, Ill., Jeffersons printing and stationery co.

Charles J Banks, Michael Chesshire, and Anne Stringfellow. 2008. "A pilot-scale trial comparing mesophilic and thermophilic digestion for the stabilisation of source segregated kitchen waste."

Cheng, Hefa, and Yuanan Hu. 2010. "Municipal solid waste (MSW) as a renewable source of energy: Current and future practices in China." *Bioresource technology* 101 (11):3816-3824.

Cho, JK., SC. Park, and HN. Chang. 1995. "Biochemical methane potential and solid state anaerobic digestion of Korean food wastes." *Bioresource Technology* 52 (3):245-253. doi: 10.1016/0960-8524(95)00031-9.

COM, European Commission. 2010. Communication from the Commission to the Council and the European Parliament on future steps in bio-waste management in the European Union. In *COM (2010) 235*. Brussels.

Conly, H. 2011. "Using DDGS as a Feedstock for Bioenergy via Anaerobic Digestion." In *Distillers Grains*, 465-474. AOCS Publishing.

Demirel, B., P. Scherer, O. Yenigun, and TT. Onay. 2010. "Production of Methane and Hydrogen from Biomass through Conventional and High-Rate Anaerobic Digestion Processes." *Critical Reviews in Environmental Science and Technology* 40 (2):116-146. doi: 10.1080/10643380802013415.

Demirel, Burak, and Paul Scherer. 2008. "The roles of acetotrophic and hydrogenotrophic methanogens during anaerobic conversion of biomass to methane: a review." *Reviews in Environmental Science and Bio/Technology* 7 (2):173-190.

Deublein, D., and A. Steinhauser. 2008. *Biogas from waste and renewable resources: an introduction*. Weinheim [Germany]: Wiley-VCH.

Deublein, Dieter, and Angelika Steinhauser. 2011. *Biogas from waste and renewable resources: an introduction*. 2nd, rev. and expanded ed. ed. Weinheim: Wiley-VCH.

Edelmann, W., K. Schleiss, and A. Joss. 2000. "Ecological, energetic and economic comparison of anaerobic digestion with different competing technologies to treat biogenic wastes." *Water Science and Technology* 41 (3):263-273.

Finnveden, Göran, Jessica Johansson, Per Lind, and Åsa Moberg. 2005. "Life cycle assessment of energy from solid waste—part 1: general methodology and results." *Journal of Cleaner Production* 13 (3):213-229. doi: http://dx.doi.org/10.1016/j.jclepro.2004.02.023.

Fujishima, S., T. Miyahara, and T. Noike. 2000. "Effect of moisture content on anaerobic digestion of dewatered sludge: ammonia inhibition to carbohydrate removal and methane production." *Water science and technology : a journal of the International Association on Water Pollution Research* 41 (3):119-127.

Fulford, David. 1988. *Running a biogas programme: a handbook*. London: Intermediate Technology Publications.

Gunnerson, C.G., and D.C. Stuckey. 1986. *Anaerobic digestion: Principles and practices for biogas systems*. Vol. 49: World Bank.

GWP. 2012. Decision No. 432/QD-TTg dated on April 12th, 2012 on "Approving the Vietnam Sustainable Development Strategy for the 2011-2020." Vietnam Government Web Portal.

Han, S.K., and H.S. Shin. 2004. "Biohydrogen production by anaerobic fermentation of food waste." *International Journal of Hydrogen Energy* 29 (6):569-577.

Hartmann, H., and BK. Ahring. 2005. "Anaerobic digestion of the organic fraction of municipal solid waste: influence of co-digestion with manure." *Water research* 39 (8):1543-1552.

He, Pin Jing. 2010. "Anaerobic digestion: An intriguing long history in China." *Waste management* 30 (4):549-550.

JICA. 2011. Study report on Solid waste management in Vietnam.

Kim, Jung Kon, Gui Hwan Han, Baek Rock Oh, Young Nam Chun, Chi-Yong Eom, and Si Wouk Kim. 2008. "Volumetric scale-up of a three stage fermentation system for food waste treatment." *Bioresource Technology* 99 (10):4394-4399. doi: 10.1016/j.biortech.2007.08.031.

Kim, SH., SK. Han, and HS. Shin. 2004. "Feasibility of biohydrogen production by anaerobic co-digestion of food waste and sewage sludge." *International Journal of Hydrogen Energy* 29 (15):1607-1616. doi: 10.1016/j.ijhydene.2004.02.018.

Kwon, Sung-Hwan, and Dong-Hoon Lee. 2004. "Evaluation of Korean food waste composting with fed-batch operations I: using water extractable total organic carbon contents (TOCw)." *Process Biochemistry* 39 (10):1183-1194. doi: 10.1016/s0032-9592(03)00233-4.

Lier, JB., N. Mahmoud, and G. Zeeman. 2008. "Anaerobic wastewater treatment." In *Biological wastewater treatment: principles, modelling and design*, 415-456. IWA publishing.

Maier, Raina M., Ian L. Pepper, and Charles P. Gerba. 2000. *Environmental microbiology*. New Delhi: Elsevier.

MONRE. 2011. Vietnam's 2011 National Environment Report - Solid waste section.

Nguyen, Hoa Huu, Sonia Heaven, and Charles Banks. 2014. "Energy potential from the anaerobic digestion of food waste in municipal solid waste stream of urban areas in Vietnam." *International Journal of Energy and Environmental Engineering* 5 (4):365-374.

Nguyen, Huu Hoa. 2014. "Modelling of food waste digestion using ADM1 integrated with Aspen Plus." Faculty of Engineering and the Environment, Univesity of Southampton.

Nguyen, Xuan Hoang, and Hoang Viet Le. 2011. "Solid waste management in Mekong Delta." *Journal of Vietnamese Environment* 1 (1):27-33.

Pertl, A., P. Mostbauer, and G. Obersteiner. 2010. "Climate balance of biogas upgrading systems." *Waste Management* 30 (1):92-99.

Rao, M. S., and S. P. Singh. 2004. "Bioenergy conversion studies of organic fraction of MSW: kinetic studies and gas yield–organic loading relationships for process optimisation." *Bioresource Technology* 95 (2):173-185. doi: 10.1016/j.biortech.2004.02.013.

Rao, P.V., S.S. Baral, R. Dey, and S. Mutnuri. 2010. "Biogas generation potential by anaerobic digestion for sustainable energy development in India." *Renewable and sustainable energy reviews* 14 (7):2086-2094.

Shin, Hang-Sik, Jong-Ho Youn, and Sang-Hyoun Kim. 2004. "Hydrogen production from food waste in anaerobic mesophilic and thermophilic acidogenesis." *International Journal of Hydrogen Energy* 29 (13):1355-1363. doi: 10.1016/j.ijhydene.2003.09.011.

Sivanagaraju, S. 2010. "Generation And Utilization Of Electrical Energy." Pearson India. http://www.myilibrary.com?id=266318.

Smyth, BM., JD. Murphy, and CM. O'Brien. 2009. "What is the energy balance of grass biomethane in Ireland and other temperate northern European climates?" *Renewable and Sustainable Energy Reviews* 13 (9):2349-2360.

Spellman, Frank R. Whiting Nancy E. 2007. *Environmental management of concentrated animal feeding operations (CAFOs)*. Boca Raton: CRC Press.

Stuckey, DC. 1986. "Biogas: a global perspective." In *Biogas Technology, Transfer and Diffusion*, 18-44. Springer.

Wall, JD., CS. Harwood, and AL. Demain. 2008. *Bioenergy*. Washington, D.C.: ASM Press.

World Bank. 2012. "Electric power consumption (kWh per capita)." The World Bank.

Zhang, Lei, Yong-Woo Lee, and Deokjin Jahng. 2011. "Anaerobic co-digestion of food waste and piggery wastewater: Focusing on the role of trace elements." *Bioresource Technology* 102 (8):5048-5059. doi: 10.1016/j.biortech.2011.01.082.

Zhang, R., HM. El-Mashad, K. Hartman, F. Wang, G. Liu, C. Choate, and P. Gamble. 2007a. "Characterization of food waste as feedstock for anaerobic digestion." *Bioresource Technology* 98 (4):929-935. doi: 10.1016/j.biortech.2006.02.039.

Zhang, Ruihong, Hamed M. El-Mashad, Karl Hartman, Fengyu Wang, Guangqing Liu, Chris Choate, and Paul Gamble. 2007b. "Characterization of food waste as feedstock for anaerobic digestion." *Bioresource Technology* 98 (4):929-935. doi: 10.1016/j.biortech.2006.02.039.

Zhu, H., W. Parker, R. Basnar, A. Proracki, P. Falletta, M. Beland, and P. Seto. 2008. "Biohydrogen production by anaerobic co-digestion of municipal food waste and sewage sludges." *International Journal of Hydrogen Energy* 33 (14):3651-3659.

BIOGRAPHICAL SCETCH

Name: Hoa Huu Nguyen

Affiliation: [1] Visiting Researcher at the University of Southampton, UK (UoS)
[2] National University of Civil Engineering, Vietnam (NUCE)

Education:

- 2010-2014: PhD Degree in Modelling of Anaerobic Digestion of food waste at the University of Southampton, UK.
- 2004-2007: Master Degree in Environmental Engineering at NUCE
- 1999-2004: Bachelor Degree in Water Engineering at NUCE

Address: Water and Environmental Engineering Group, University of Southampton, UK

University Road
Southampton
SO17 1SG

Research and Professional Experience:

Dr Hoa Nguyen has been working on research projects in Vietnam and in the UK since 2004. His projects were about water supply and drainage, wastewater treatment and water resource management. Currently, his research interests are: modelling of anaerobic digestion of various types of organic feedstocks using Matlab and Aspen Plus simulation software; modelling of ammonia removal from digestate; and other renewable energy topics.

Professional Appointments:

Dr Hoa Nguyen is currently working for a company specialised in designs and builds biogas plants throughout the UK. However, he is still a lecturer at National University of Civil Engineering, Vietnam. He is also working as a visiting researcher at the University of Southampton in research projects on modelling of anaerobic digestion processes.

Publications Last 3 Years:

1. Energy potential from the anaerobic digestion of food waste in municipal solid waste stream of urban areas in Vietnam. International Journal of Energy and Environmental Engineering, 2014.
2. Simulation and optimisation of single stage biogas upgrading and bottling system using Aspen Plus. International Conference on Anaerobic Digestion (Biogas Science), Austria, 2014.
3. Energy balance of food waste anaerobic digestion. EU-China Workshop on Resource Recovery from Biomass and Green Technology, 2014.

In: Food Waste
Editor: Garrett Leonard Riley
ISBN: 978-1-63485-025-4

Chapter 6

MEASURING FOOD WASTE GENERATION FROM HOUSEHOLDS IN GREECE: BEHAVIOURS, ATTITUDES, AND POTENTIAL FOR PREVENTION

K. Abeliotis**, *C. Chroni and K. Lasaridi
Harokopio University, Athens, Greece

ABSTRACT

Food waste is among the priority streams for waste prevention worldwide, as it is generated in large quantities at all levels of the food production and consumption chain, in both developed and developing countries, throughout the year. Previous research indicates that food waste produced by households is one of the most significant parts of the overall food waste stream, at least in developed countries. Quantifying food waste, as well as investigating the causes of the ever-growing waste generation throughout the lifecycle of food chain, brings out the environmental, economic and social impacts of food waste. At the household level, food waste is associated with certain behaviours, that need to be identified and comprehended, in order to prevent the generation of food waste. This chapter overviews the generation of food waste within households and investigates the potential for prevention,

*Corresponding Author address Email: kabeli@hua.gr.

showcasing a questionnaire-and-diary-based study, which was the first of its kind in Greece.

Keywords: food waste behaviours, waste prevention, households, Greece

INTRODUCTION

In early 2010s the Food and Agricultural Organization (FAO) estimated that approximately one third of food produced for human consumption is either lost or wasted through the food supply chain, from agricultural production and post-harvest handling and storage to processing, transportation, distribution and consumption (Gustavsson et al. 2011). According to a study for food waste in the EU-27, the retail-wholesale sector produces approximately 5% of the total amount of food waste, the manufacturing sector 39%, households 42%, while food service - catering 14% (EC 2011). Besides the association of food waste generation with environmental impacts, such as waste generation and climate change, it is also associated with economic impacts, such as business and consumers savings (Nahman et al. 2012), in addition to social impacts such as the global food and water security (Koivupuro et al. 2012; Kommu et al. 2012) or the expanding dimensions of malnutrition, in both undernourishment (Nahman et al. 2012) and obesity terms (Parfitt et al. 2010).

From a worldwide point of view, the contribution of each sector differs substantially due to a range of reasons dependent on the socio-economic conditions of each country. Parfitt et al. (2010) stated that in developing countries food losses are higher at the immediate post-harvest stages, while in industrialised and developing economies they are higher for perishable foods after they have been prepared for market, or, in fact, sold to the consumer; "*for affluent economies, post-consumer food waste accounts for the greatest overall losses*." A recent analysis of the food supply sectors by region, showed that wastage occurring at consumption level is much more variable, with substantial wastage in middle- and high-income regions, but much lower in low-income regions (FAO 2013).

Back in 2009, WRAP estimated that consumers in the UK throw away 31% of the food they purchase for consumption (WRAP 2009). There is growing evidence that the contribution of the households to the food waste problem is particularly significant (Sharp et al. 2010a; EC 2011). In EU-27, households are responsible for 42% of the total amount of food waste generated (EC 2011). Food waste at the consumption and household level is becoming an increasingly significant global issue and food waste mitigation is becoming a priority.

In order to tackle food waste generation in households a number of socio-demographic, behavioural and attitudinal factors that could have an impact on food waste reduction have been studied (Evans 2012a; Koivupuro et al. 2012; Stefan et al. 2013; Abeliotis et al. 2014; Stancu et al. 2016; Visschers et al. 2016). Thyberg and Tonjes (2016) reviewed the drivers of food waste in an effort "*to provide insight into the best policy approaches to sustainably manage food waste*." Koivupuro et al. (2012) showed that the amount of food waste generated in Finnish households is statistically correlated with the size and the type of household, the gender of the person who has the main responsibility for shopping, people's view on the waste prevention potential, the provision of food products in low prices, and the household's view on the impact of purchasing different package sizes. Understanding such behaviours is crucial for planning effective campaigns for food waste reduction: groups of people to focus on, channels of engagement and design of campaign material (Tucker 2007; WRAP 2009). From a sociological point of view, Evans (2011; 2012b) argued that "*household food waste cannot be conceptualised as a problem of individual consumer behaviour*," but sould be rather seen as a matter of social and material (e.g., "*material infrastructure of provision*") conditions in which food is provisioned.

Among the different waste prevention initiatives that can be implemented by waste-stream at the household level, food waste ranks as the top stream, with the highest prevention potential (Cox et al. 2010; Salhofer et al. 2008).

This chapter reviews the evidence on the amounts of food waste produced by households in Greece and their causes.

Definitions

One of the barriers in the study and quantification of food waste generation and prevention is the lack of a universally accepted definition (Thyberg and Tonjes 2016). In this chapter food waste is defined as "*composed of raw or cooked food materials and includes food loss, before, during or after meal preparation in the household, as well as food discarded in the process of manufacturing, distribution, retail and food service activities*" (EC 2011), while food loss "*refers to a decrease in mass (dry matter) or nutritional value (quality) of food that was originally intended*" but is no more suitable "*for human consumption*" (FAO 2013). Food wastage comprises both food waste and loss. However, Schneider (2013) indicated that there is no fully agreement on a definition of food waste as regards prevention. In order to pass this obstacle, food waste in literature is quite often split into two categories, regarding their prevention potential: avoidable and non-avoidable food waste. Waste prevention measurements aim at the reduction of the former (Schneider 2013).

Behaviours and Attitudes of Households towards Food Waste

At the household level, food waste generation is strongly associated with socio-economic parameters (age, income, marital status) as well as with certain attitudes and behaviours, such as meal planning, cupboard checking and list making before going for shopping, proper storage of food items, use of food leftovers, cooking the right amount of food and good understanding of expiration date labels (WRAP 2009; Abeliotis et al. 2014; Koivupuro et al. 2012; Stefan et al. 2013; Quested et al. 2013). However, studies also indicate that parameters of food provision, such as packaging, also affect the generation of food waste (Langley et al. 2011; Evans 2012b; Williams et al. 2012). In order to prevent the generation of food waste, those behaviours must be identified and comprehended. Understanding these behaviours can also help to plan campaigns providing insides on: which groups of people to focus on; channels of engagement; and design of campaign material (WRAP 2009).

There is a large number of behaviours that can have a positive impact on food waste generation and, consequently, to the efforts towards food waste prevention (Tucker 2007; WRAP 2009). Such behaviours include meal planning, cupboard checking and list making before going for food shopping, proper storage of food items, use of food leftovers, cooking the right amount of food and careful use of expiration date labels (Quested et al. 2013). As already mentioned, food waste is an important field of waste prevention activities with high potential. The study of behavioural aspects towards waste prevention has appeared recently in the literature: for instance, Bortoleto et al. (2012) developed and tested a hybrid model for the prediction of household waste prevention. Focusing on food waste, Stefan et al. (2013) applied the Theory of Planned Behaviour (TPB) for the study of food waste prevention by the consumers in Romania. The authors concluded that Romanian consumers' food waste generation is mainly driven by routines related to food provisioning rather than by an intention not to waste food. The same theory was applied by Karim Ghani et al. (2013) in a completely different geographical setting, in Malaysia, to study the influencing factors of participation in source separation of food waste. Koivupuro et al. (2012) studied the influence of certain socio-demographical, behavioural and attitudinal factors on the amount of avoidable food waste generated in Finnish households. The authors showed that in Finnish households the amount of food waste is correlated with the size of the household, the gender of the person who is responsible for grocery shopping, the frequency of buying discounted products, the view of the respondent to food reduction and the habit of purchasing food packet sizes (Koivupuro et al. 2012). Visschers et al. (2016) considered the TPB to be a useful tool for the prediction of *"consumer's food waste behaviour because the latter is under volitional control."*

In spite of the indisputable value of the abovementioned pioneering studies on waste prevention behaviour, as well as the studies of WRAP (2009), Koivupuro et al. (2012) and Schneider and Lebersorger (2009) on the patterns and causal factors of food waste generation, more research is needed, in order to guide efficient campaigns to promote households engagement in food waste prevention initiatives.

METHODOLOGY

The research took place via the use of a structured questionnaire in the Athens metropolitan area (representing an urban centre, i.e., with population >10,000 inhabitants) and in the cities of Larissa, Trikala and Volos in the region of Thessaly in central Greece (representing semi-urban centres, i.e., population <9.999 inhabitants) during February and March 2012. The use of questionnaires for conducting food waste related studies is well documented (Koivupuro et al. 2012; Stefan et al. 2013; Schneider and Lebersorger 2009; WRAP 2009). Two hundred and fifty five (255) participants were randomly selected and completed the questionnaire handed to them which was later collected by members of the research team. Of those, 231 fully completed the questionnaire. Thus the response rate was 90.6%. Only people who were involved in food purchases and cooking in their households were asked to fill the questionnaire. This was further verified by placing a screening question in the introductory section of the questionnaire. The results of the questionnaires provide quantitative information, such as kind and frequency of behaviours and habits that result in wasted food and reasons for those habits on the basis of self-reported behaviour of the respondents. As Bortoleto et al. (2012) stated, self-reported behaviour can be used as a proxy, given the impossibility of observing and reporting prevention behaviours of all the participants in a survey. Furthermore, information about interrelated conditions (e.g., level of employment, age of household members) and behaviour (e.g., buying, cooking and diet habits) can be gathered. Attitude and behaviour surveys are commonly used in waste prevention efforts (Kurisu and Bortoleto 2011; Sharp et al. 2010b; Zorpas and Lasaridi 2013); they are used to set a baseline at first and later to evaluate changes in attitude or behaviour (Sharp et al. 2010b), despite the fact that this baseline is 'declared' and at times deems subjective (Sharp et al. 2010a).

RESULTS AND DISCUSSION

The demographic information of the sample is given in Table 1. The two columns on the left present the characteristics of the household; columns on the right present the characteristics of the respondents, respectively. The majority of the respondents are mostly between 25-54 years of age (77.4%); they hold a university (48.7%) or a secondary education degree (26.1%); they live in apartment buildings (71.9%) that they own (65.7%). Thus, the people that participated in the research are young and well-educated. The chi-square test indicated significant differences with the general urban population in Greece concerning these two parameters, but the sample is representative within the aforementioned groups. In terms of the household location, approximately three out of four households are located in urban areas.

Table 1. The demographics of the sample

Household characteristics	%	Age	%
Single person	16.5	18-24	10.9
Couple	23.8	25-34	25.6
Family (w/ young children)	30.7	35-44	25.7
Family (all adults)	29.0	45-54	26.1
		>55	11.7
Home characteristics	%	Educational background	%
Single family home	28.1	Compulsory education	6.9
Apartment building	71.9	Secondary education	26.1
Own home	65.7	University	48.7
Rent home	34.3	Postgraduate studies	18.3

Previous studies indicated that there is a significant effect of food waste generation and the person who is mainly responsible for food purchasing and cooking in each household (Koivupuro et al. 2012). Thus in the beginning of the questionnaire there were two screening questions: 91.8% of the respondents are exclusively or mainly responsible or equally share the responsibility for the food supplies of their households. Moreover, 83.1% of them are exclusively or mainly responsible or equally share the responsibility for cooking in their households. The questionnaires of those people that responded that they do not participate in either food purchasing or cooking were excluded from the analysis.

The first question focused on identifying the level of annoyance of the respondents regarding the generation of food waste. The question was *"To which extent are you concerned or bothered about the amount of food waste that your household throws away?"* 53.4% replied "a great deal," 36.4% replied "a fair amount," 5.1% "a little" and only 5.1% replied "not very much." The reported high level of annoyance is very promising since it indicates that Greek households exhibit high awareness on food waste. Comparable high awareness results regarding food waste is also reported for citizens from the region of Lower Austria (Schneider and Lebersorger 2009). However, Stefan et al. (2013) found that just the intention not to waste food does not have a significant effect on reported food wastage. On the contrary, food waste is a food-related behaviour and it therefore depends on culture; thus the generation of food waste is a result of the daily routines that consumers perform and effective prevention of food wastage may not be the result of conscious intentions not to waste food (Stefan et al. 2013). This is further supported by the qualitative findings of Evans (2012), according to which respondents in the UK also appeared not to hold a callous or careless disregard for the food that they waste; nevertheless food wastage is substantial. Thus, beyond the level of annoyance towards food waste, consumer habits expressed as reported behaviour have to be identified.

Based on the original food waste related behaviours identified by WRAP, nine different behaviours that affect food waste generation were compiled for Greece:

- Food waste behaviour
- Meal planning
- Check food stocks
- List making
- Use of freezer

- Storage of open cheese/meat
- Use of fridge to store fruit/vegetables
- Use of leftovers
- Cooking the right amount
- Knowledge of date labels

As Quested et al. (2013) explain, such behaviours are proxies for a wider range of behaviours. Tracking these behaviours over time provides an indication of behaviour change, rather than an absolute measure (Quested et al. 2013). The nine behavioural scales examined are analysed in the following lines:

Before going for food shopping, in order to reduce food waste, it is good to perform the following three actions: (i) check the cupboards at home; (ii) make a menu plan for the days that follow immediately; and (iii) write down a list of the food items that need to be purchased. The engagement in these three behaviours was reported by the respondents using a 5-point Likert-type scale. The choices of the scale were termed as "Always," "Very often," "Sometimes," "Rarely," "Never." Meal planning is the most frequently practiced behaviour compared to list making and checking the food stocks. According to Stefan et al. (2013), planning routines, such as meal planning and consistent list making, have a negative effect on food waste generation, while shopping routines are positively associated with food waste.

Storing of food under sub-optimal conditions leads to losses in quality and therefore increased wastage (Koivupuro et al. 2012). Therefore, the next behaviour studied is about the use of refrigerators for the storage of food in each household: more specifically, the way that five items (meat, cheese, apples, carrots and food leftovers) are stored was checked. Fresh products are much more vulnerable to decay. Thus, their storage is the refrigerator is necessary. Based on this fact, the next two scales are focusing more specifically on the refrigeration of (i) meat and cheese and (ii) of fruits and vegetables. Proper storage of meat and cheese is performed much more frequently compared to the refrigeration of fruits and vegetables.

Not re-using leftovers is among the most common reasons for food wastage (Koivupuro et al. 2012). Thus, the next behavioural scale deals with the handling of food leftovers. Two questions constitute this scale: (a) "*What do you normally do with the leftovers in your plate?*" and (b) "*What do you normally do with the leftovers left in your cooking pot?*" If a respondent selected that he/she "keeps them for a next meal," then they were asked to reply about the final fate of the kept leftovers: The responses were "always/almost always," "sometimes," or "rarely/never." If a respondent answered that he/she "keeps the leftovers for a next meal" and they "always consume them," three points were added in the scale. If they "consume them sometimes" one point was added. Finally, if they "throw them away" after refrigeration, no point was credited for this behaviour. Thus, if a person reported that he/she kept the leftovers from both his/her plate and cooking pot and always or almost always consumed them, scored the highest score. If he/she did not keep the leftovers neither from the plate or the cooking pot, he/she scored the lowest score. In this way, the respondents were only credited for reusing the leftovers and not when they just refrigerated them. Refrigeration of food, as Evans (2012) cynically explained, is often the middle step before its wastage, a way to ameliorate anxieties and guilt of the members of the household. The results indicate that the frequency of proper use of leftovers for human consumption is slightly above the mean value of the scale). However this reported behaviour presents very high standard deviation, as a result of the multiplicity of people's behaviours. The use of leftovers is a food waste prevention behaviour which is contradictory to the social finding that several authors have noted: food waste is an outcome of food-related behaviour; this means that people who provide food for their households are more interested in the freshness or the nutritional quality of food items (Stefan et al. 2013; Evans 2012a). As a result, food that does not (or is not perceived to) meet the aforementioned criteria, including leftovers, ends up as waste.

Preparing or serving greater portions of food often yields to more food wasted. In order to address this behaviour, the next scale dealt with the amount of food prepared: the question was straightforward: "How often do you consider that the amount of food that you cook is the right one?" Again, the engagement in this behaviour was reported by the respondents using a 5-point Likert-type scale. The choices of the scale were termed as "Always," "Very often," "Sometimes," "Rarely," "Never." Approximately 60% of the respondents replied that they were always or very often thinking that the amount of food that they cook in not the right one. In a similar survey in Austria, Schneider and Lebersorger (2009) reported that 27% of the respondents admitted that they often cooked more food than they could eat, but the majority of the respondents (89%) used these food residues later for other meals.

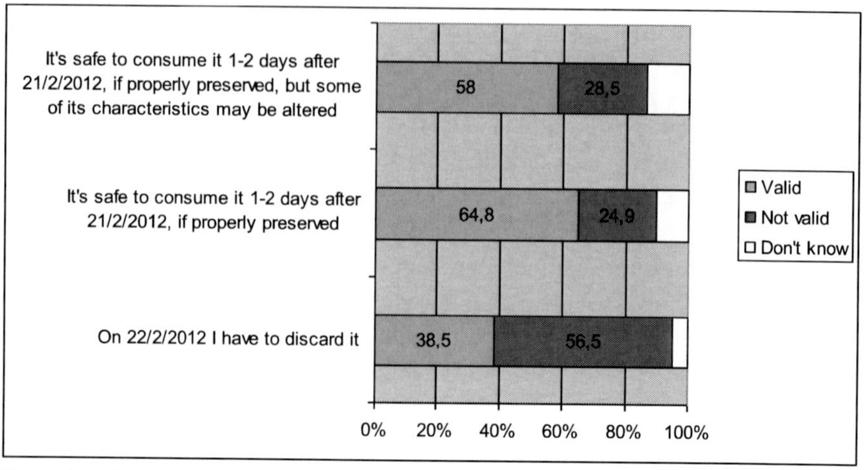

Figure 1a. "If you read on a food label that it is best before 21/2/2012, which of the following is valid?"

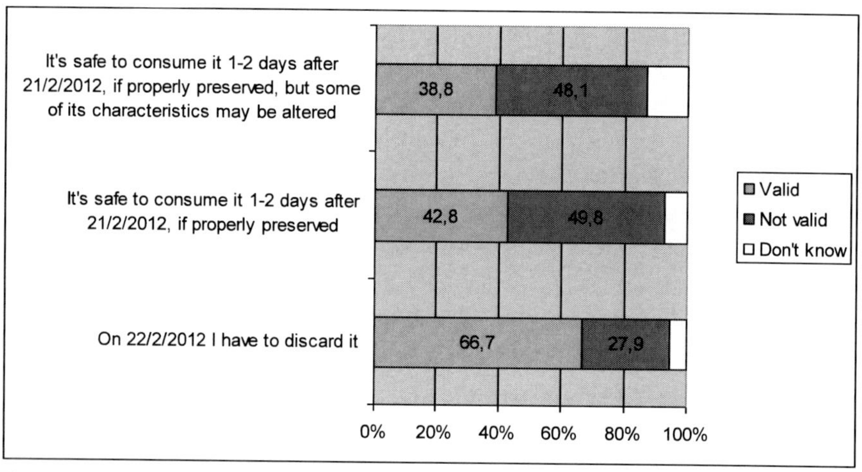

Figure 1b. "If you read on a food label that the item is expiring on 21/2/2012, which of the following is valid?"

Another source of food waste is that of products past their "expiration" or "best used before" dates. Correct knowledge of the meaning of those two labels is required by the consumers in order to prevent unnecessary food wastage. The "expiration" date is a stricter regulatory limit compared to the "best used before" date. Again, the actual nature of the food item may have an effect on the consumer's awareness on the item's expiration or "best used before" labelling. The frequency of checking the various food storage places within the household, such as cupboards, refrigerator and freezer, determines the percentage of food wasted. Frequent checking can yield to better management of food items closer to their expiration or "best used before" dates and consequently to lower amounts of food waste. In order to identify the level of knowledge, three definitions were given to the consumers for each one of the labels; only one among them was the correct one. The definitions and the respective responses of consumes are illustrated in Figures 1a and 1b. Fifty-eight percent (58.0%) of the respondents replied correctly about the "best used before" label while 66.7% replied correctly about the definition of the "expiration date" label. Comparable results (56%) for the knowledge of "best used before" were reported for consumers in Austria by Schneider and Lebersorger (2009). However, in our case, the reported knowledge of food date labels is slightly below the mean value of the scale, which indicates that there is plenty of room for education of the consumers regarding the proper meaning of the two food date labels.

Finally, the influence of the adverse economic situation in Greece on the behaviour of the households was also examined. The first question in this part of the questionnaire asked the respondents to compare the current amount of food that is thrown away with the respective amount 12 months ago. Interestingly enough, none of the respondents replied that the food waste has increased during the past 12 months. On the contrary, 32.0% replied that the amount remained the same, 25.1% replied that the food thrown way decreased a little, while 42.9% replied that the amount thrown away decreased a lot due to the current financial situation in Greece. The next question aimed at identifying the actions that yielded the difference in the amount of food that is thrown away, as reported in the previous question (see Table 2). The most practiced action was list making (48.0%), followed by buying only the food that is absolutely necessary (45.5%) and checking the cupboards before going for shopping (44.2%).

Table 2. "Which of following actions yielded the reduction of food waste that is thrown away during the past 12 months?"

	%
I only buy food that is absolutely needed	45.5
I prepare a weekly meal schedule	37.2
I try to buy food stuff, only after I consume whatever I have at home	43.6
I started cooking the leftovers from previous days	17.3
I make better estimates the food portions cooked and served	28.2
I buy more local food products	13.5
I care more for the "best used before" and "expiration date" labels	9.0

Table 2. (Continued)

	%
I use my freezer more	22.4
I make a list before going for food shopping	48.0
I check the cupboards at home before going for food shopping	44.2

Table 3. How often do you perform the following actions now compared to 12 months ago?

Actions	Current frequency (times per week)	Frequency 12 months ago (times per week)	SD
Using the leftovers for creative cooking	2.4	-	-
Eating out or having food delivered	0.8	1.96	0.908
Eating frozen pre-cooked food	0.4	2.39	0.976
All family members have meal together	4.8	3.57	0.959
Invitation from relatives or friends	0.8	2.86	1.006
Take food from parents	1.0	2.98	0.908
Cooking for other family members not living with you	0.8	2.74	0.943

A third question in this part explored the change in the frequency of certain behaviours during the past 12 months. The respondents were asked to place their preference on a 5-point Likert-type scale ranging from "much more frequent" to "much more rarely." Much more frequently was expressed as 5 while "much more rarely" was expressed as 1 in the Likert scale. Table 3 presents the comparison in the frequency of the given options compared to those 12 months ago. It is evident that the frequency of eating out, ordering food and buying pre-cooked food has been significantly reduced because of the financial recession, since all of the aforementioned behaviours are directly related to the available income of the households. On the positive social side, the frequency of family members sharing the same meal has been increased.

CONCLUSION

Food waste is a stream with multiple social, economical and environmental implications generated throughout the life cycle stages of food. Among those stages, in developed countries, such as Greece, the role of the households in the generation of food waste is very important. Therefore, in order to demonstrate its prevention potential, quantification of the household food waste is required, so that the key sources of its generation are identified.

The present research was undertaken as part of an ongoing project which aims at measuring food waste generation in Greece. A study on nine behaviours of households related to food waste has been conducted for the first time in Greece. Behaviour surveys are used, at first, to set a baseline; then, they are used to evaluate changes in behaviour. Thus, the current study could serve as a baseline before the introduction of prevention measures and policies aiming at the reduction of food waste generation in Greece. On the negative side, the major limitation of our study is that the nine scales measuring the dependent variables contained self-reported items which may be biased estimates of true behaviour.

In terms of the proper behaviour of consumers towards the reduction of food waste, three simple things have to be promoted: at first, consumers should buy the right amount of food; then, they should store and preserve the food bought in the best possible way; and finally, the consumers should use all the food that is bought in the right amount and properly preserved.

Also, in order to prevent the generation of food waste, the knowledge on the "expiration date" and "best used before" food labels should be improved. In addition, the cooking skills of those involved in cooking should also be improved. This could also be supported by the preparation of dedicated cookbooks with recipes that make easier the use of leftovers.

Overall, results of this study, to our knowledge the first of its kind in Greece, can be of great assistance to the effective development of food waste prevention campaigns. Such campaigns can produce positive environmental and economic results, at both the household level and the overall waste management system in Greece. A specific challenge would be to build on the current, necessity driven change and demonstrate its environmental and moral virtues, to create a "feel good" factor among the population and sustain good practices and pro-prevention behaviour upon economic recovery.

Effective actions are underway by international organisations, central and local authorities in various countries, as well as the private sector, setting precedents for worldwide adaptation and replication. Among them, some initiatives focus on restraining unsustainable consumption patterns. Yet, much more work needs to be done, engaging the main principles of sustainable production, supply, consumption and discard patterns and waste management. Moreover, there are still several barriers and challenges, such methodological limitations, conflicts of interest and established human attitudes and behaviours that need to be resolved. An effort to approach waste prevention targets, considering social and market patterns, might hold a key to the design of effective waste prevention measures and the move to the - so needed - resource efficient societies.

ACKNOWLEDGMENTS

The study on food waste generated by households in Greece was supported by the LIFE10 ENV/GR/622 project "Development and Demonstration of a Waste Prevention Support Tool for Local Authorities – WASP Tool," which was co-funded by the European Union LIFE+ programme.

REFERENCES

Abeliotis, K., Lasaridi, K., Chroni, C., 2014. "Attitudes and behaviour of Greek households regarding food waste prevention." *Waste Management and Research* 32, 237 - 240.

Bortoleto A. P., Kurisu K. H., Hanaki K., 2012. Model development for household waste prevention behaviour. *Waste Management* 32(12), 2195-2207.

Cox. J., Giorgi S., Sharp V., Strange K., Wilson D. C., Blakey N., 2010. Household waste prevention – a review of evidence. *Waste Management and Research* 28: 193-219.

EC, 2011. Preparatory study on food waste across EU-27. Technical Report - 2010-054. European Communities ISBN: 978-92-79-22138-5, DOI: 10.2779/85947.

Evans D., 2011. Blaming the consumer – once again: the social and material contexts of everyday food waste practices in some English households. *Critical Public Health* 21, 429-440.

Evans D., 2012a. Binning, gifting and Recovery: The conduits of disposal in household food consumption. *Environment and Planning D: Society and Space*, 30, 1123-1137.

Evans D., 2012b. Beyond the throwaway society: Ordinary domestic practice and a sociological approach to household food waste. *Sociology* 46, 41-56.

FAO, 2013. Food wastage footprint, Impacts on natural resources. Summary report. Food Wastage Footprint – FAO.

Gustansson J., Cederberg C., Sonesson U., van Otterdijk, Meybeck, 2011. Global Food Losses and Waste. Extent, Causes and Prevention. A study conducted for the International Congress SAVE FOOD! From FAO, Rome 2011: http://www.fao.org/docrep/014/mb060e/mb060e00.pdf (February 2016).

Karim Ghani W. A. W. A. K., Rusli I. F., Biak D. R. A., Idris A., 2013. An application of the theory of planned behaviour to study the influencing factors of participation in source separation of food waste. *Waste Management* 33, 1276-1281.

Koivupuro H.-K., Hartikainen H., Silvennoinen K., Katajajuuri J.-M., Heikintalo N., Reinikainen A., Jalkanen L., 2012. Influence of socio-demographical, behavioural and attitudinal factors on the amount of avoidable food waste generated in Finnish households. *International Journal of Consumer Studies* 36: 183-191.

Kummo M., de Moel H., Porkka M., Siebert S., Varis O., Ward P. J. (2012). Lost food, wasted resources: Global food supply chain losses and their impacts on freshwater, cropland, and fertiliser use. *Science of the Total Environment* 438, 477-489.

Kurisu K. H., Bortoleto A. P., 2011. Comparison of waste prevention behaviours among three Japanese megacity regions in the context of local measures and sociodemographics. *Waste Management* 31: 1441-1449.

Langley J., Turner N., Yaxall A., 2011. Attributes of packaging and influences on waste. *Packaging Technology and Science* 24, 161- 175.

Lebersorger S., Scheider F., 2011. Discussion on the methodology for determining food waste in household waste composition studies. *Waste Management* 31, 1924-1933.

Nahman A., de Lange W., Oelofse S., Godfrey L., 2012. The costs of household food waste in South Africa. *Waste Management* 32: 2147-2153.

Parfitt J., Barthel M., Macnaughton S., 2010. Food waste within food supply chains: quantification and potential for change to 2050. *Philosophical Transactions of the Royal Society* 365: 3065-3081.

Quested T. E., Marsh E., Stunell D., Parry A. D., 2013. Spaghetti soup: The complex world of food waste behaviours. *Resources, Conservation, Recycling* (2013), http://dx.doi.org/10.1016/j.resconrec.2013.04.011.

Salhofer S., Obersteiner G., Schneider F., Lebersorger S., 2008. Potentials for the prevention of municipal solid waste. *Waste Management* 28, 245–259.

Schneider F., 2013. Review of food waste prevention on an international level. *Waste and Resource Management* 166, 187-203.

Schneider F., Lebersorger S., 2009. Households Attitudes and Behaviour towards Wasting Food – A Case Study. In: Cossu et al. (Eds.), Proceedings of 12th Sardinia 2009 Symposium, pp. 831-832.

Sharp V., Giorgi S., Wilson D. C, 2010a. Delivery and impact of householdwasteprevention intervention campaign (at the local level). *Waste Management and Research* 28: 256-268.

Sharp V., Giorgi S., Wilson D. C., 2010b. Methods to monitor and evaluate household waste prevention. *Waste Management and Research* 28: 269-280.

Stancu V., Haugaard P., Lahteenmaki L., 2016. Determinants of consumer food waste behaviour: Two routes to food waste. *Appetite* 96, 7 – 17.

Stefan V., van Herpen E., Tudoran A., Lähteenmäki L., 2013. Avoiding food waste by Romanian consumers: The importance of planning and shopping routines. *Food Quality and Preference* 28, 375–381.

Thyberg K. L., Tonjes D. J., 2016. Drivers of food waste and their implications for sustainable policy development. *Resources, Conservation and Recycling* 106, 110 – 123.

Tucker, P., Douglas, P., 2007. Understanding Household Waste Prevention Behaviour. A project for Defra's WREP.

Visschers V. H. M., Wickli N., Siegrist M., 2016. Sorting out food waste behaviour: A survey on the motivations and barriers of self-reported amounts of food waste in households. *Journal of Environmental Psychology*, 45, 66 -78.

Williams H., Wikström F., Otterbring T., Löfgren M., Gustafsson A., 2012. Reasons for household food waste with special attention to packaging. *Journal of Cleaner Production* 24, 141-148.

WRAP, 2009. Household Food and Drink Waste in the UK. Report prepared by WRAP. Banbury. www.wrap.org.uk/foodanddrinkwaste (April 15, 2012).

Zorpas A. A., Lasaridi K., 2013. Measuring waste prevention. *Waste Management* 33(5), 1047-1056.

In: Food Waste
Editor: Garrett Leonard Riley

ISBN: 978-1-63485-025-4

Chapter 7

SAFETY ANALYSIS OF SOY SAUCE SQUEEZING RESIDUE AND FISH MEAL STORAGE BY THERMAL ANALYSIS AND GAS CHROMATOGRAPHY

Naoharu Murasawa*
Faculty of Risk and Crisis Management,
Chiba Institute of Science, Choshi, Japan

ABSTRACT

An increased demand for recycling has prompted the food industry to become more efficient in its handling of waste. However, there are some cases in which food wastes or products manufactured by recycling have spontaneously ignited during transport or storage, or else have depleted oxygen in their storage area, causing the deaths of workers. This study investigated heat generation and oxygen levels during the storage of soy sauce squeezing residue and fish meal, by-products of soy sauce and fish production and processing, respectively, by thermal analysis and gas chromatography. Results suggest that oxygen deficiency may occur in a

* Corresponding author: Naoharu Murasawa. Faculty of Risk and Crisis Management, Chiba Institute of Science, 3 Shiomi-Cho, Choshi, 288-0025, Japan. E-mail: n-murasawa@cis.ac.jp.

well-sealed storage facility due to fermentation. However, if oxygen is continuously cycled and the facility stores large deposits in an insulated environment, fermentation can cause the temperature to increase, leading to the oxidation of fatty acid esters and, eventually, fire.

Keywords: soy sauce squeezing residue, fish meal, fermentation, fire forensics, thermal analysis

1. Introduction

Recent trends aim at increasing the environmental sustainability of the modern human lifestyle in order to alleviate the burden of society on the environment. In Japan, this tendency has seen additional support since the Great East Japan Earthquake – indeed, the Japanese government has increasingly focused on replacing fossil fuels and nuclear energy with alternative energy sources, with particular focus being placed on biomass fuels [1]. This approach takes advantage of the 3Rs principle (reduce, reuse, and recycle), the basis for sustainable waste management in Japan and elsewhere; in this way, techniques for implementing the policy continuously improve over time [2]. However, even with advances in recycling research, recycled materials can still pose a safety hazard during production, distribution, and storage if they are not fully assessed for potential dangers [3, 4].

Soy sauce and fish are important ingredients in Japanese cuisine. These two ingredients generate waste products that are referred to as soy sauce squeezing residue (SSSR) and fish meal. The food industry has begun to use these waste products as fuel; SSSR, for example, can be incinerated, generating thermal energy. However, SSSR and fish meal can spontaneously ignite during storage and transport [5]. In addition, SSSR has caused oxygen deficiency in storage areas due to fermentation, reportedly resulting in the death of workers [6].

Unfortunately, since the composition of waste material is not fixed, it can be challenging to simulate the appropriate conditions for safety testing. Furthermore, it is considerably more difficult to determine potential hazards and identify the driving forces in heat generation, both in the early stages of a fire and after a fire accident. Therefore, thermal analysis is often used to investigate the cause of fire, wherein a small sample is used to study the properties of a material to help prevent fire, large quantities of poisonous gas, and smoke in future incidents.

This risk assessment can also focus primarily on materials that are likely to generate heat when piled. In this study, various thermal analysis methods were applied to the hazard assessment of SSSR and fish meal, specifically to examine the environments leading to spontaneous ignition and oxygen deprivation.

2. Materials and Methods

2.1. Samples

Large amounts of SSSR are produced when unrefined soy sauce, or moromi, is compressed during the raw soy sauce extraction process. SSSR is plate shaped when produced but is later finely ground for storage and processing (Figure 1). Fish meal, meanwhile, is obtained by boiling residual fish substances to remove water and oil, then pressing it and drying the remaining solid with hot air or steam; the resulting material is a crushed powder (Figure 2). In this experiment, some samples of both SSSR and fish meal were subjected to a 17 h sterilization treatment using ethylene oxide gas (EOG) in order to minimize fermentation [7].

Figure 1. SSSR.

Figure 2. Fish meal.

2.2. Thermogravimetric Differential Thermal Analysis (TG-DTA)

A TG-DTA system (Rigaku Thermoplus TG 8120, Japan) was used to study the overall thermal characteristics of the samples (Figure 3). The samples (each approximately 20 mg) were placed in an open aluminium container (0.05 ml) and were then heated from room temperature to 600°C at a rate of 2 K/min. Air was circulated at 150 ml/min.

2.3. Calvet Calorimetry

A Calvet calorimeter (Setaram C80, France) was used for additional thermal testing. The C80 is a highly sensitive, twin-type heat-flux calorimeter (Figure 4) that reduces the effects of residual water evaporation given that it contains the sample in a high-pressure closed vessel (8 ml). This also allows it to take measurements from room temperature to 100°C, which is a temperature range that cannot be easily measured using TG-DTA. A 1500-mg sample was used at a heating rate of 0.1 K/min, beginning at room temperature and increasing up to 300°C.

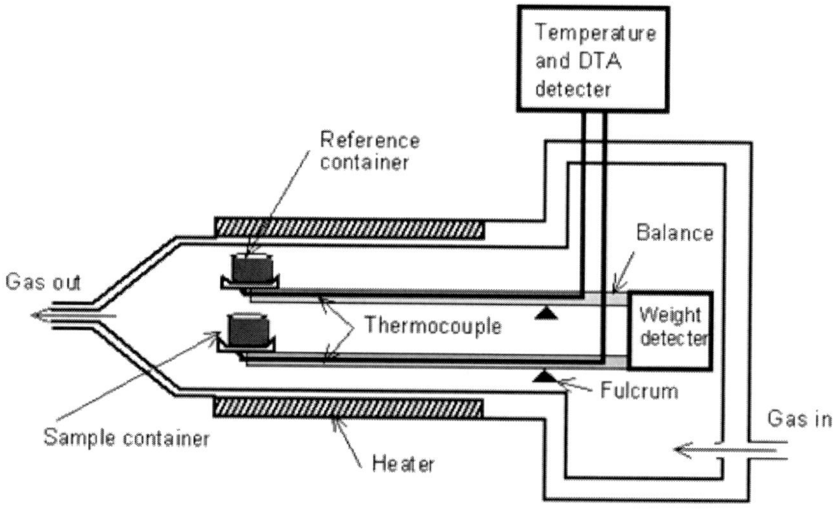

Figure 3. Schematic diagram of the TG-DTA system.

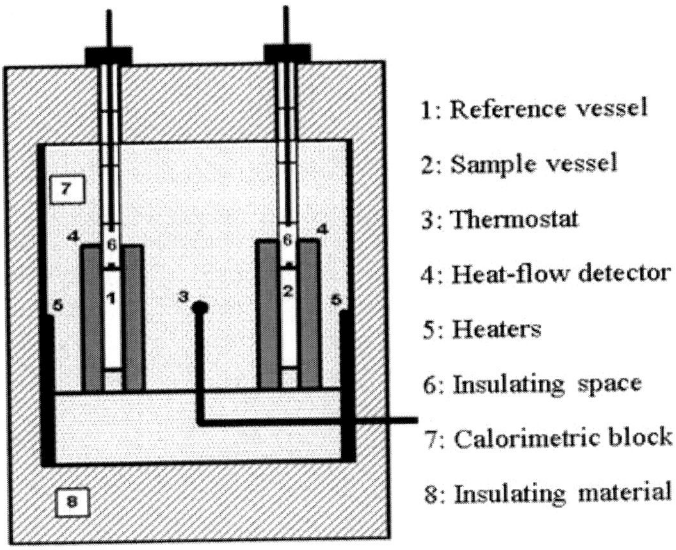

Figure 4. Schematic diagram of the Calvet calorimeter.

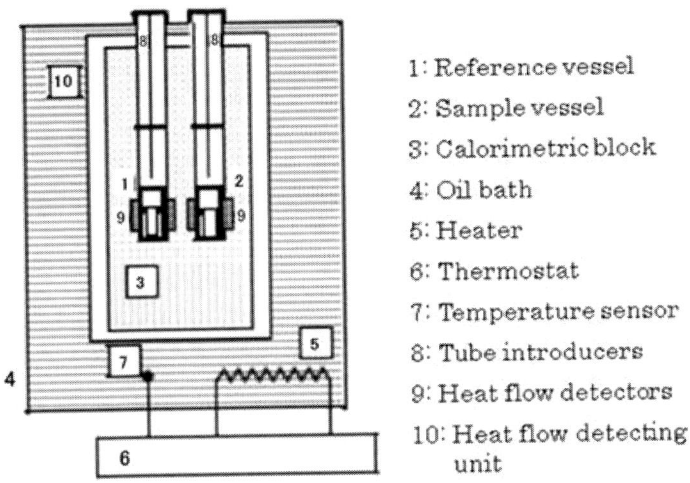

Figure 5. Schematic diagram of the TAM thermometric calorimeter.

2.4. High-Sensitivity Isothermal Calorimetry

A highly sensitive isothermal calorimeter (Thermometric TAM-III, Sweden) was used to examine faint heat generation from the fermentation and oxidation of fatty-acid esters in detail. The TAM (Figure 5) can measure the amount of heat generated by microbial fermentation at the nanoscale. A sample of 1000 mg was placed in a sealed container (4 ml) that was maintained at 50°C for 3 days.

2.5. Gas Chromatography (GC)

To study gas emission during storage, 50 g samples were placed in 1-L glass bottles that were sealed and placed in a thermostatically controlled oven which kept the bottles at temperatures over a range of 5-50°C for 10 days. The resulting gas was collected (Figure 6) and measured by GC (Shimadzu GC-14B, Japan) with a standard gas (CO: 0.05%; C_2H_6: 0.995%; H_2: 0.097%, CO_2: 0.996%; CH_4: 0.987%) for calibration and a thermal conductivity detector (200°C; sensitivity: 50 mA; carrier gas: Ar, 20 ml/min). The column temperature increased from 40°C (6 min hold) to 80°C (12 min hold), and then finally to 150°C (10 min hold), at a rate of 40°C/min. An air cylinder (O_2: 21%; N_2: 79%) was used as the standard gas for O_2 and N_2 measurements. Detector sensitivity was changed to 30 mA for this measurement, while the isothermal column temperature was maintained at 30°C.

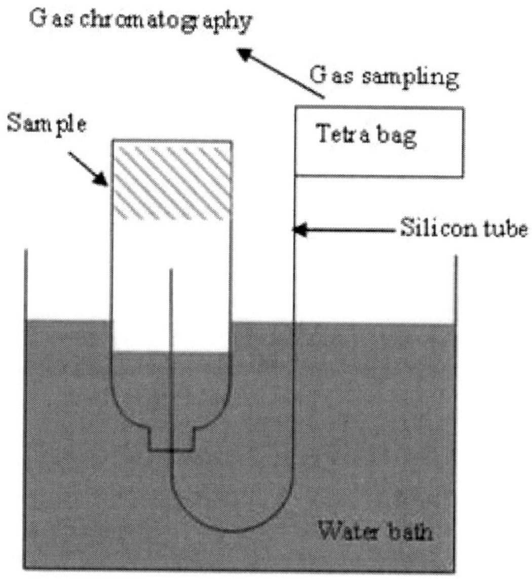

Figure 6. Schematic of the GC measuring device.

3. Results and Discussion

3.1. TG-DTA

The TG-DTA results are shown in Figure 7, while the percent mass losses at 100°C and decomposition temperatures are shown in Table 1. The decomposition temperatures are the points at which the DTA curve shifted by 0.1 μV (0.01K) in the heat generation direction from the constant baseline. In the DTA curve, the downward direction indicates an endothermic reaction while the upward direction indicates an exothermic reaction.

The thermal decomposition of SSSR and fish meal is divided into three phases. Because of SSSR dehydration, a total weight decrease of 26% was observed from room temperature to 100°C. Next, organic components decomposed and combusted from 180 to 380°C. By the end of this phase, a total weight loss of about 70% was observed. The final stage was consistent with the decomposition and combustion of carbide, leaving a residue of just 10% of the original.

Meanwhile, fish meal showed a 7.8% dehydration loss from room temperature to 100°C.

Overall, fish meal lost mass at a lower rate than SSSR, resulting in a smaller heat generation peak but in greater remaining residue. A shift to heat generation occurred at 172.9°C and 187.3°C for SSSR and fish meal, respectively. This is, theoretically, the point at which combustion would begin.

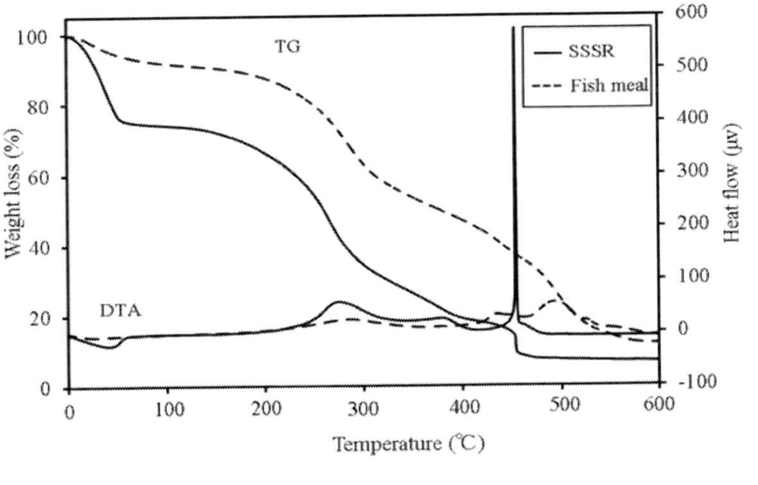

Figure 7. TG-DTA results for SSSR and fish meal.

Table 1. TG-DTA mass loss at 100°C and decomposition temperature

Sample	Mass loss at 100°C (%)	Decomposition temperature (°C)
SSSR	25.6	172.9
Fish meal	7.8	187.3

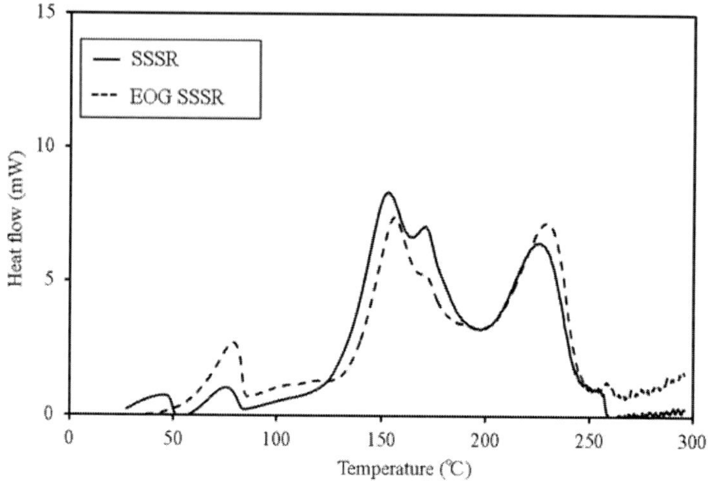

Figure 8. Calvet calorimetry results for SSSR and SSSR treated with EOG.

3.2. Calvet Calorimetry

Figures 8 and 9 show the SSSR and fish meal calorimetry results, respectively, demonstrating the effect of EOG sterilization in each case. Table 2 summarises the exothermic onset temperatures of each sample. The exothermic onset temperature is the temperature at which the rate of heat generation increases to 0.005 mW.

In untreated SSSR, heat is generated immediately after the measurement begins. In comparison, the EOG-treated SSSR sample generates no heat at this temperature, suggesting that fermentation causes the observed generation of heat in the untreated sample. Although heat generation in untreated SSSR decreases at approximately 50°C, immediately after it begins in EOG-treated SSSR, it then recovers, mirroring behaviour at lower temperatures. The untreated fish meal, meanwhile, begins to generate heat at a slightly lower temperature than the treated sample. However, little difference is observed overall below 50°C, while a very similar pattern is observed above that point.

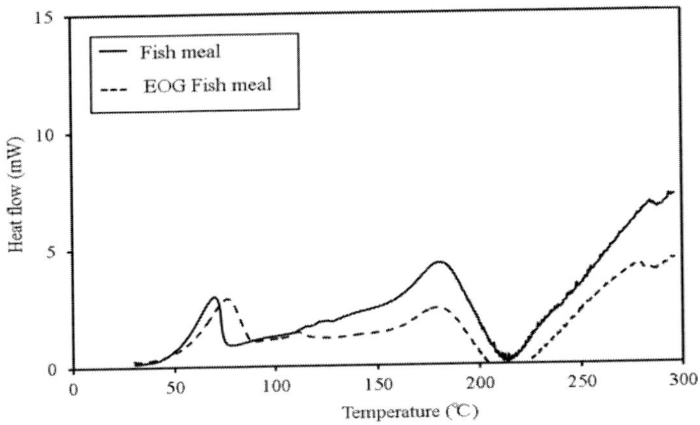

Figure 9. Calvet calorimetry results for fish meal and fish meal treated with EOG.

Table 2. Heat generation onset temperature of sterilized and non-sterilized SSSR and fish meal

Sample	Heat generation onset temperature (°C)
SSSR	27.6
EOG SSSR	53.9
Fish meal	41.2
EOG Fish meal	42.3

In general, the heat generated from the oxidation of fatty acid esters is observed over a temperature range of 80-100°C [8]. Beyond 100°C, heat generation is assumed to be caused by decomposition [9-11]. Overall, fermentation plays a greater role in SSSR, while fatty acid esters play a greater role in the fish meal.

3.3. High-Sensitivity Isothermal Calorimetry

The SSSR and fish meal TAM measurements are shown in Figures 10 and 11, respectively. Table 3 summarizes the amount of heat generated over three testing periods: 0-24 h, 24-72 h, and 0-72 h. The temperature was set to 50°C because microbial activity peaks at approximately this temperature [12-13] and gradually decreases at higher temperatures. This test was performed both in a limited air atmosphere and under nitrogen.

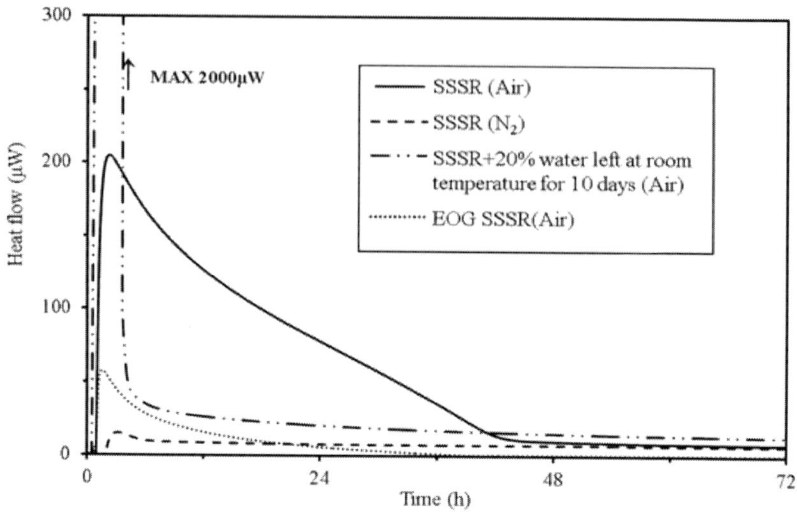

Figure 10. TAM results for SSSR.

The amount of microbial activity and, therefore, heat generated by microbial fermentation, the primary cause of said heat generation, can be further increased by adding water to the sample and allowing it to sit for a prolonged period. Accordingly, the test was performed with approximately 20% water added to 10 g of sample and maintaining 1 g of the sample at 25 ± 5°C for 10 days.

No significant heat generation was observed in a nitrogen atmosphere for either sample. However, in air, heat generation spiked immediately for both samples, then gradually decreased. Although SSSR heat generation peaked higher, the difference between the total amounts of heat generated by the two samples was not significant. The addition of water yielded a sharp exothermic peak (2000 µW) for the SSSR that was not observed with the fish meal. Overall though, each reaction slowed, with the more active ones doing so much more quickly. This suggests the rapid consumption of oxygen early on.

On the basis of these results, it seems that heat generation from microbial fermentation plays a more prominent role in SSSR than in fish meal; this may be due to the fact that the higher moisture content of SSSR makes it more suitable for microbial proliferation.

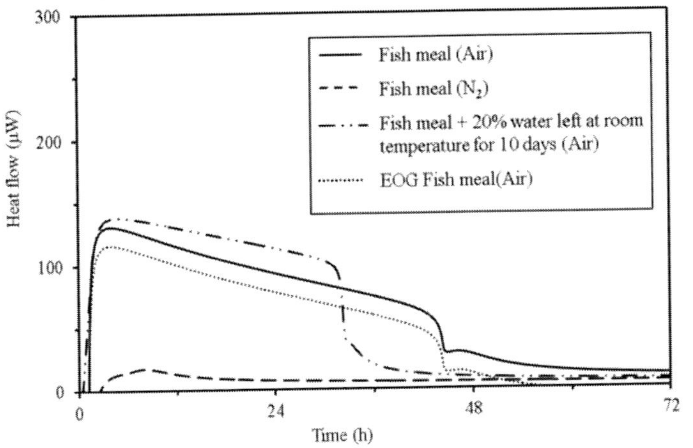

Figure 11. TAM results for fish meal.

3.4. GC

GC was used to analyse the fermentation gases and determine the effect of lower temperatures on the process. As fermentation was affected by moisture content, several samples also had 20% distilled water added. The results are shown in Tables 4 and 5, while a summary of oxygen and carbon dioxide concentrations in relation to the holding temperature is shown in Figures 12-15.

Carbon dioxide accounts for the majority of gas produced during SSSR and fish meal storage. Small amounts of flammable gases, specifically carbon monoxide and hydrogen, were also detected from fish meal. Furthermore, an increase in the production of carbon dioxide was observed when moisture was added in both cases, with the overall amount increasing up to 30°C; after this point, it begins to decrease with temperature.

With SSSR, oxygen content frequently dropped below the maximum acceptable limit for human safety. This applied even at storage temperatures below 25°C, especially for high moisture content. This danger is only ever present for fish meal stored above 25°C. Because fermentation increases temperature during storage [14-15], sufficient attention must be paid to the humidity and ventilation conditions.

Table 3. Heat generation at 50°C

Sample	Heat generation (J/g) 0–24 h	Heat generation (J/g) 24–72 h	Heat generation (J/g) 0–72 h
SSSR(Air)	11.1	4.2	15.3
SSSR(N2)	0.7	1.2	1.9
SSSR+ distilled water 20% left at room temperature for 10 days (Air)	14.4	2.5	16.9
EOG SSSR(Air)	2.6	1.3	3.9
Fish meal(Air)	8.2	7	15.2
Fish meal(N2)	0.9	1	1.9
Fish meal+ distilled water 20% left at room temperature for 10 days (Air)	11.6	5.3	16.9
EOG Fish meal(Air)	10.6	4.9	15.5

3.5. Combustion Mechanism

Figure 16 summarizes the different circumstances that can trigger spontaneous ignition or oxygen deficiency at storage facilities for fermentable food wastes. The C80 and GC results demonstrate that fermentation likely increases the generation of heat and production of carbon dioxide up to approximately 50°C. Likewise, the C80 and TAM results indicate a transition to heat generation by fatty acid ester oxidation at temperatures above 50°C. When water evaporates and temperatures exceed 100°C, the wastes begin to thermally crack and, possibly, combust. The fact that oxygen deficiency is most likely to occur in sealed storage facilities with little circulation requires that oxygen be actively introduced; however, since this could lead to the combustive processes described above, special care must be taken in planning storage facilities.

CONCLUSION

SSSR has a greater moisture content than fish meal, accounting for its greater heat generation due to fermentation. However, fish meal also exhibited heat generation due to fatty acid ester oxidation. This heat generation can begin even at low temperatures, suggesting that appropriate care must be taken in avoiding build-up and fire.

Table 4. GC SSSR results

Sample	Storage period	Storage temperature	GC analysis results (%)					
			O_2	N_2	H_2	CO	CH_4	CO_2
SSSR	10 days	5 ℃	18.6	77.4	-	-	-	1.3
SSSR +Distilled water20 %			12.3	77.8	-	-	-	5.8
SSSR		15 ℃	16.2	78.8	-	-	-	2.4
SSSR +Distilled water20 %			5.3	79.3	-	-	-	15.3
SSSR		25 ℃	1.7	79.4	-	-	-	16.1
SSSR +Distilled water20 %			1.7	79.1	-	-	-	18.2
SSSR		30 ℃	1.7	81.2	-	-	-	16.6
SSSR +Distilled water20 %			1.5	79.6	-	-	-	18.2
SSSR		40 ℃	2.1	79.2	-	-	-	15.3
SSSR +Distilled water20 %			0.9	80.3	-	-	-	16.4
SSSR		50 ℃	12.8	80.2	-	-	-	5.6
SSSR +Distilled water20 %			10.2	78.9	-	-	-	7.5

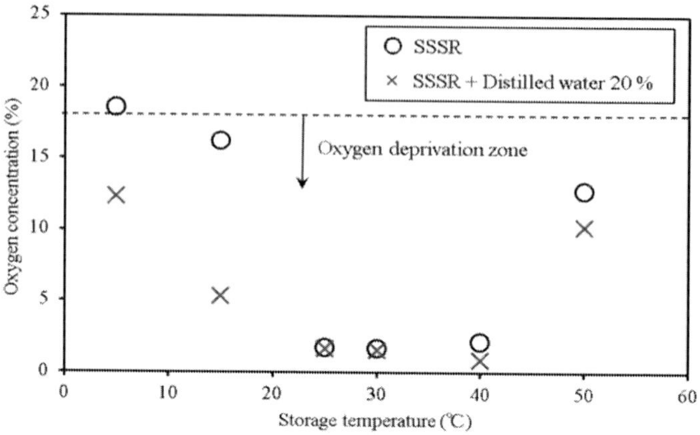

Figure 12. Correlation between storage temperature and oxygen concentration of SSSR and SSSR with 20% distilled water.

Table 5. GC fish meal results

Sample	Storage period	Storage temperature	GC analysis results (%)					
			O_2	N_2	H_2	CO	CH_4	CO_2
Fish meal	10 days	5 ℃	20.4	76.7	-	-	-	0.04
Fish meal +Distilled water20 %			19.6	77.1	-	-	-	0.13
Fish meal		15 ℃	20.3	76.8	-	-	-	0.04
Fish meal +Distilled water20 %			18.7	77.5	-	-	-	0.6
Fish meal		25 ℃	20.1	77.4	0.001	0.003	-	0.5
Fish meal +Distilled water20 %			4.0	83.3	0.001	0.01	-	11.6
Fish meal		30 ℃	14.3	81.3	0.003	0.05	-	0.9
Fish meal +Distilled water20 %			3.5	82.1	0.006	0.007	-	12.3
Fish meal		40 ℃	16.1	79.7	0.003	0.04	-	0.6
Fish meal +Distilled water20 %			5.7	80.5	0.03	0.07	-	10.6
Fish meal		50 ℃	17.2	79.2	0.006	0.1	-	0.5
Fish meal +Distilled water20 %			6.9	80.2	0.007	0.2	-	8.5

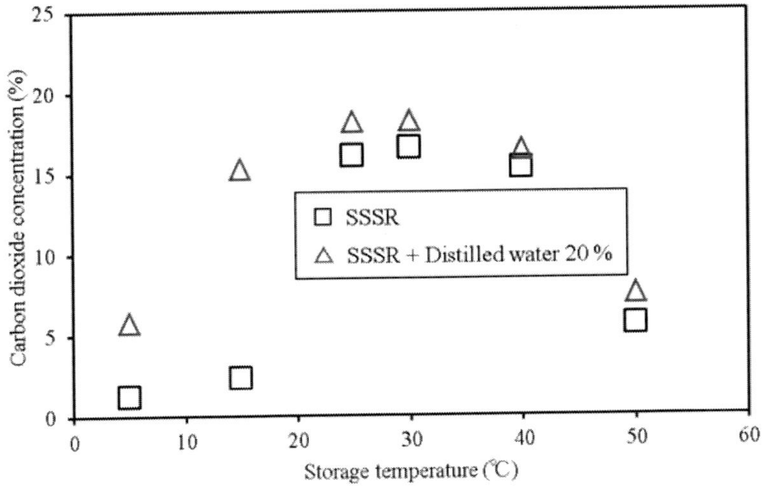

Figure 13. Correlation between storage temperature and carbon dioxide concentration of SSSR and SSSR with 20% distilled water.

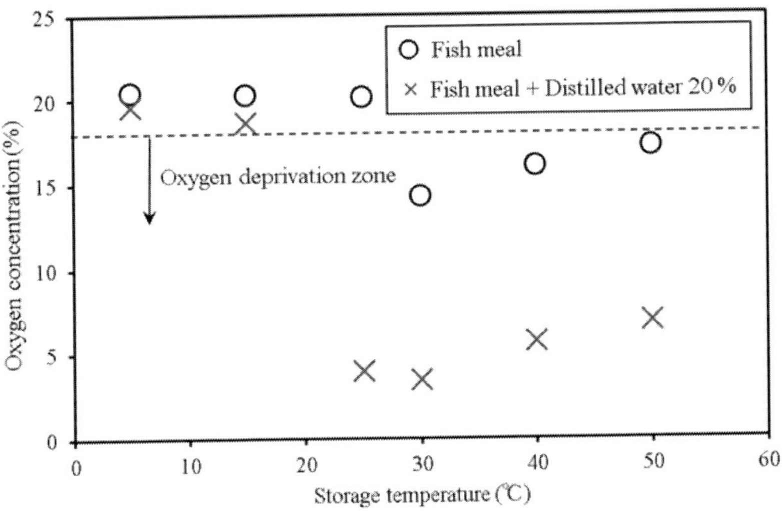

Figure 14. Correlation between storage temperature and oxygen concentration of fish meal and fish meal with 20% distilled water.

Restricting the access of moisture and oxygen to the waste itself can in part counter this by preventing significant fermentation, partly accounting for the biggest concern in the more dangerous storage of SSSR. Isolating waste and preventing access to the air also avoids the danger of oxygen depletion through fermentation, a significant safety risk at room temperature for SSSR and at only slightly elevated temperatures for fish meal. If this is not possible for large waste deposits, the use of gas meters can help to track the build-up of flammable degradation gases like methane and carbon monoxide. Overall, these results should help to improve safety conditions in food waste storage facilities, both generally and for those containing SSSR and fish meal in particular.

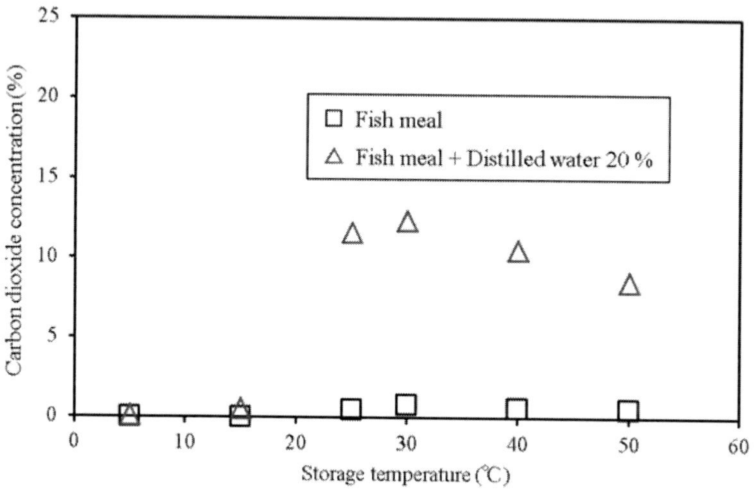

Figure 15. Correlation between storage temperature and carbon dioxide concentration of fish meal and fish meal with 20% distilled water.

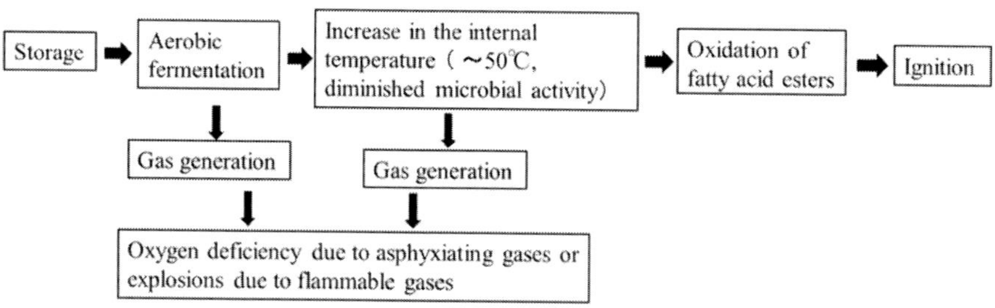

Figure 16. Causes of spontaneous ignition and oxygen deficiency.

REFERENCES

[1] K. Takeshi. "Review of energy and climate policy developments in Japan before and after Fukushima," *Renewable and Sustainable Energy Reviews*, Vol. 43, pp. 1320-1332 (2015).

[2] G. John and Y. Kaya. "Greenhouse Gas Control Technologies - 6th International Conference." Proceedings of the 6th International Conference on Greenhouse Gas Control Technologies, 4 October 2002, Kyoto, Japan (2003).

[3] Z.M. Fu, X.R. Li, H. Koseki. "Heat Generation of Refuse Derived Fuel with Water," *Journal of Loss Prevention in the Process Industries*, Vol. 18, Issue 1, pp. 27-33 (2005).

[4] A. Yasuhara, Y. Amano, T. Shibamoto. "Investigation of the Self-Heating and Spontaneous Ignition of Refuse-Derived Fuel (RDF) During Storage," *Waste Management*, Vol. 30, Issue 7, pp. 1161-1164 (2010).

[5] N. Murasawa, H. Koseki, Y. Iwata, L. Gao. "Study on spontaneous ignition of stored food waste to be used for recycling," *Fire and Materials*, Vol. 37, Issue 7, pp. 520-529 (2013).

[6] N. Murasawa, H. Koseki, Y. Iwata, Y. Shibata. "Determination of Spontaneous Ignition of SSSR and Fish Meal during Transport and Storage," *Journal of Food Research*, Vol. 1, pp. 320-329 (2012).

[7] E. H. Gillespie, J. M. Jackson, G. R. Owen. "Ethylene oxide sterilisation--is it safe?," *Jounal of Clinical Pathology*, 32, pp. 1184-1187 (1979).

[8] X.R. Li, H. Koseki, M. Momota. "Evaluation of danger from fermentation-induced spontaneous ignition of wood chips," *Journal of Hazardous Materials*, 135, no. 1-3, pp. 15-20 (2006).

[9] B.F. Gray, J.F. Griffiths, S.M. Hasko. "Spontaneous ignition hazards in stockpiles of cellulosic materials: criteria for safe storage," *Journal of Chemical Technology and Biotechnology*, 34, no. 8, pp. 453-463 (1984).

[10] J. Rychlý, L. Matisová-Rychlá, M. Lazár, K. Slovák, M. Strlič, D. Kočar, J. Kolar. "Thermal oxidation of cellulose investigated by chemiluminescence. The effect of water at temperatures above 100 C," *Carbohydrate Polymers*, 58, no. 3, pp. 301-309 (2004).

[11] S. Gaan, P. Rupper, V. Salimova, M. Heuberger, S. Rabe, F. Vogel. "Thermal decomposition and burning behavior of cellulose treated with ethyl ester phosphoramidates: Effect of alkyl substituent on nitrogen atom," *Polymer Degradation and Stability*, Vol. 94, Issue 7, pp. 1125-1134 (2009).

[12] N. Murasawa, H. Koseki, Y. Iwata, T. Sakamoto. "Lessons learned from fires of the Rubble- Risks of spontaneous ignition caused by the microbial activity," *Loss Prevention Bulletin*, Issue 226, pp. 16-19 (2012).

[13] N. Murasawa, H. Koseki. "Investigation of heat generation from biomass fuels," *Energies*, Vol. 8, Issue 6, pp. 5143-5138 (2015).

[14] R. K. Thauer, K. Jungermann, K. Decker. "Energy conservation in chemotrophic anaerobic bacteria," *Bacteriological Reviews*, Vol. 41, Issue 1, pp. 100-180 (1977).

[15] Q. Wang, J.-Y. Narita, W. Xie, Y. Ohsumi, K. Kusano, Y. Shirai, H. Ogawa, "Effects of anaerobic/aerobic incubation and storage temperature on preservation and deodorization of kitchen garbage," *Bioresource Technology*, Vol. 84, Issue 3, pp. 213-220 (2002).

In: Food Waste
Editor: Garrett Leonard Riley

ISBN: 978-1-63485-025-4

Chapter 8

SUGARCANE BAGASSE HEMICELLULOSE PROPERTIES, EXTRACTION TECHNOLOGIES AND XYLOOLIGOSACCHARIDES PRODUCTION

Michel Brienzo[1,*], Ana Flavia Azevedo Carvalho[2], Franciane Cristina de Figueiredo[3] and Pedro de Oliva Neto[2]

[1]Laboratory of Biomass Characterization,
Bioenergy Research Institute (IPBEN),
Universidade Estadual Paulista (UNESP), Rio Claro-SP, Brazil
[2]Laboratory of Development of Bioprocess,
Bioenergy Research Institute (IPBEN) Assis,
Universidade Estadual Paulista (UNESP), Assis-SP, Brazil
[3]Departament of Biochemistry and Microbiology, Universidade
Estadual Paulista (UNESP), Rio Claro, Brazil

ABSTRACT

The sugarcane has been used for centuries for sugar production and, in the recent decades, for ethanol fuel production through biotechnological routes. From sugar and ethanol industry a large amount

* Corresponding author: mbrienzo@ipben.unesp.br.

of waste is generated as a by-product, the sugarcane bagasse (SCB). This waste is one of the largest lignocellulosic material (LCM) resources, characterized by its chemical composition based on the main components of cellulose, hemicellulose and lignin. SCB is burned for energy co-generation; however, several applications have been studied based on its components, e.g., hemicellulose has been studied for the production of biofuel (xylose fermentation to ethanol), organic acids, artificial sugar (xylitol), and derivatives such as xylooligosaccharides (XOS). In spite of its several important applications, the hemicellulose extraction or solubilization from LCM is not simple. Hemicellulose is associated with cellulose and lignin in the plant cell wall in a highly organized structure. The LCM is recalcitrant and resistant to biodegradation or biotechnological conversion routes. The LCM requires a pretreatment step to change the accessibility of the polysaccharides or these need to be extracted for further uses. The hemicellulose can be extracted based on chemical (acidic or alkaline treatment) and enzymatic processes, which depends on the final application (mono, oligo or polysaccharides). XOS are special oligomers with potent prebiotic effect, usable as additives for food and feed. However, they are poorly exploited due to the cost and technological limitations. In this review were discussed the SCB hemicelluloses properties and the key technologies involved in the hemicellulose extraction and XOS production. Additionally, studies that prove the effectiveness of the use of these oligomers as a prebiotic for human and animal were presented from recent reports.

Keywords: sugarcane bagasse, hemicellulose, xylan, alkaline extraction, xylanases hydrolysis, acid hydrolysis, autohydrolysis, xylooligosaccharides, prebiotic effect

INTRODUCTION

The research of new alternatives for energy and food is a global concern due to technological, physical and environmental limitations. There has been an increase in agricultural waste production because of an intensive population growth and thus this knowledge is required for food and fuels. The problem begins with the inappropriate disposal of generated wastes that can provoke environmental problems. Furthermore, these agricultural wastes can be used as a feedstock material to biotechnological processes, in order to produce energy and added-value molecules. To use waste material so as to convert it into a valuable product a pretreatment or a process for biomolecule extraction is necessary. Despite several decades of study, there are still some questions that need to be answered about LCM. These specific wastes from agriculture and forest are recalcitrant and resistant to enzyme action, characterized by their heterogeneity (Brienzo et al., 2014).

Sugarcane bagasse (SCB) is currently one of the most important lignocellulosic wastes generated from sugar and ethanol production. Brazil is the largest producer, responsible for 40% of the worldwide production. The sugarcane is cultivated in Brazil to produce sugar (since many centuries ago) and ethanol and electricity (in the last decades) in an integrated plant. The SCB is the waste generated after sugarcane is crashed for sucrose extraction. From the sugarcane harvesting another waste is generated: the straw, of which part is left on the field for soil fertilization. Both of the LCM, bagasse and straw, are materials rich in fiber and show potential to bioenergy or macromolecule extraction. The composition of these materials is based on cellulose, hemicellulose and lignin. In brief, these macromolecules are interesting for several added value products, for instance, cellulose: used in fermentation to solvent and biofuels and cellulose-derived products such as carboxymethyl cellulose and cellulose nitrate; hemicellulose: used in fermentation to solvent and biofuels, biofilm, xylooligosaccharides; lignin: phenols, activated carbon, carbon fiber. Among several possible applications of sugarcane waste, the use in electricity generation is predominant, not considering its biotechnological potential.

Among biomass components, the hemicellulose is a special class of polysaccharides characterized by their diversity and heterogeneity. The hemicellulose is wide dispersed among plants, with different polysaccharides on hardwood (glucuronoxylan and glucomannan) and softwood (galacto-glucomannan and arabino-glucuronoxylan) (Fengel and Wengels, 1984). For SCB, the hemicellulose has been reported as mainly composed by xylan polysaccharides. Furthermore, the most common hemicellulose polysaccharides are xylans, constituted of xylose units. The structure of the xylan content is highly dependent on the plant source and also can vary according to the sugarcane plant fractions (Brienzo et al., 2014). The materials rich in xylan are highly interesting to XOS production. The XOS can be produced from LCM using chemical methods, such as autohydrolysis, steam explosion, acid catalyzed medium, and enzymatic hydrolysis or a combination of these kinds of methods (Brienzo et al., 2010; Yang et al., 2005; Vázquez et al., 2000).

Xylooligosaccharides are not digestible in the human stomach and they are considered as prebiotic as they stimulate beneficial microorganisms in the human guts. The XOS are non-cariogenic and can be used as dietary sweeteners as low-calorie diet food. The XOS are ingredients of functional foods, providing health benefits (Vázquez et al., 2000). Therefore, these sugars can improve the modulation of the colonic microbiota, especially bifidobacteria and lactobacilli (Vázquez et al., 2000; Nabarlatz et al., 2007; Gullón et al., 2008). Hence, the XOS are beneficial to the organism improving the bowel's function, calcium absorption, protection against cardiovascular diseases and also decreasing the risk of colon cancer, due to the formation of smaller fatty acids chains (Grootaert et al. 2007; Wang et al. 2009). Effects related to skin and blood, immunological action, anti-oxidant activities, anti-inflammatory and antiallergenic effects are also noted (Aachary and Prapulla, 2009). Despite all these health benefits, the XOS are poorly produced and consumed by humans and animals, because their cost is higher than other products in this same category. Therefore, more research is necessary to make the technology of XOS production more available and economical, especially when relating to prebiotic XOS.

Recently, the structure and health benefits of XOS have been reviewed (Carvalho et al., 2013). This review emphasizes the physiological properties and benefits for the body, such as health improvement through nutrient adsorption increase and factors related to inhibiting pathogenic microorganisms and consequences of a good gut development. XOS properties are dependent on its structure, which relies on the kind of sugars present and also on the number of sugars units (degree of polymerization - DP). Xylan is a polysaccharide that can be highly branched (acetyl groups, arabinose, galactose and feruloyl groups).

The market and consumption of oligosaccharides (including XOS) are currently gradually increasing. Nowadays, Japan is one of the biggest producers and consumers of oligosaccharides, with relevance to Asian countries. Furthermore, it is expected that, with the increase of health consciousness by the humans, the market of products with oligosaccharides will also increase. The human consciousness can be interpreted here as the importance and role of functional foods (added oligosaccharides) in the improvement of human health. Moreover, replacing the animal antibiotic uses for functional food should be considered as a serious issue, considering the negative effects of these compounds. The effect of antibiotic residues in meat for human consumption is not actually clear, but it is known that the overuse of antibiotics causes pathogens strains to increase (Carvalho et al., 2013). Considering the oligosaccharides as functional food ingredients of big importance and market prediction, there is a great interest in XOS production. Currently, researchers are dedicated to make the processes of production of

XOS cheaper, including studies to understand the physiological function and properties of XOS.

Focusing on the SCB as a resource of hemicellulose (xylan) and as feedstock for XOS production, the following discussion reviewed the chemical composition and properties of the SCB and its xylan. The key technology involved in the extraction of xylan and XOS production was discussed. The effectiveness and properties of these oligosaccharides were presented as well as their use as a prebiotic for humans and animals.

SUGARCANE BAGASSE AS XYLAN SOURCES

Sugarcane is a C4 (a reference to a pathway in which CO_2 is first fixed into a compound that contains four carbon atoms) perennial sucrose-storing grass, belonging to the genus *Saccharum* which has its origins in Asia. Sugarcane is cultivated in tropical and subtropical countries throughout the world, including the bigger producers Brazil (721 million ton), India (347 million ton), China (123 million ton), and Thailand (96 million ton). The SCB is generated after the crashing process conducted for sucrose extraction. The sugarcane culm is based on internode (sucrose storage) intercalated by short nodes (transversal septa). These tissues are different on what refers to physicochemical properties, vascular bundle distribution and response to pretreatment (Brienzo et al., 2014). The SCB contains two types of fiber: the cellulose fiber of rind, and the pith of the stem. The pith fiber is composed mainly of parenchyma cells and the rind of sclerenchyma cells (Miller et al., 2012). These structures distributed in the SCB make the material heterogeneous and have different recalcitrance to biotechnological routes conversion.

The SCB is a fibrous material that corresponds to 15% of the sugarcane (dry basis). The chemical composition of SCB is based on the macromolecule cellulose, hemicellulose and lignin, with a low amount of extractives and ash (Table 1). Cellulose forms microfibrils compounding a network structure with hemicelluloses, encrusted by lignin. This structure is the base of the plant cell wall, which is highly organized and make the isolation of the macromolecules more difficult. The chemical composition of the cell wall can vary according to the type of biomass and plant fraction, with differences in the polysaccharides organization (Carpita and Gibeaut, 1993).

Cellulose is recognized as one of the most abundant organic compounds in the world. Approximately 35-50% of the dry mass of SCB is cellulose.

Cellulose is a linear homopolysaccharide, composed by β-D-anhydro-glucopyranose units, which are linked by β(1,4)-glycosidic linkages. Cellulose (and hemicellulose) is a structural polysaccharide of the plant cell wall, what justifies its resistance and defiance on separating it from hemicellulose and lignin. Cellulose chains are organized with intra- and intermolecular hydrogen bonds, forming microfibrils structures, what characterizes the highly ordered (crystalline) or less ordered (amorphous) regions (Fengel and Wengels, 1984). The cellulose content is higher in the external fraction of the sugarcane culm, comprising 50%, the internode contains 44% and the node 42% of cellulose (Brienzo et al., 2014). The cellulose content in SCB varies according to species (Table 1).

Hemicelluloses are the second most important polysaccharides in biomass. Hemicelluloses consist in short chains of branched heteropolysaccharides, composed of hexoses and pentoses. The main constituents of the pentoses are D-xylose and L-arabinose. The main constituents of the hexoses are D-glucose, D-galactose and D-mannose. The sugars D-manose, L-rhamnose and D-galactose are more scarcely reported and appear in a small amount in SCB xylan (Peng et al., 2009). The major hemicelluloses component of SCB is xylose-based. Hemicellulose comprises 22-36% of the SCB, defined as L-arabino-(4-O-methyl-D-glucurono)-D-xylan (Sun et al., 2004). The type and amount of hemicelluloses vary widely, depending on plant and its tissue type. SCB fractions have a different amount of hemicelluloses: 16, 27 and 29% of bagasse mass on the external fraction, node and internode, respectively (Brienzo et al., 2014). Xylan is the representative hemicelluloses of SCB since its composition is more than 80% of xylose in the backbone (Rocha et al., 2015; Brienzo et al., 2009). The xylose content in SCB has been reported as a range of 8.8 to 20.4%, to both classical and precision breeding SCB varieties (Benjamin et al., 2013).

Lignin is the second most abundant organic (aromatic) compound, after cellulose, found in the plant. Lignin is responsible for joining the cellulose fibers with the hemicelluloses network, providing rigidity to the cell wall. The lignin physical distribution surrounding the fibers protects the polysaccharides from chemical or physical extraction and microbial attack. This characteristic enhances the cell wall recalcitrance, making it difficult to isolate hemicellulose. Lignin is an amorphous macromolecule, defined as a three-dimensional polymer with a phenylpropane structure. The lignin structure is formed by enzyme-mediated radical coupling of the three monolignols that, in the molecule, are named p-hydroxy-phenylpropane (H), syringil (S) and guaiacyl (G) units. The ratio between lignin units is suggested to be related to the material recalcitrance; the S unit is desirable to generate material with a lower resistance to the delignification process (Chiang and Funaoka, 1990). Lignin has been accepted to be responsible for the biomass recalcitrance and studies have shown that a good strategy is to select SCB feedstock with low lignin content for a better response to pretreatment and enzymatic digestibility (Brienzo et al., 2015). Lignin content varies according to the species in a range of 14-30% (Table 1). SCB fractions have a different amount of lignin: 29% in the external fraction and 22% in the internode and node (Brienzo et al., 2014). This difference of lignin content among these fractions could imply a different amount of hemicellulose extraction from each fraction.

The cell wall has other components that are not chemically linked to the cellulose-hemicellulose-lignin matrix, the extractive compounds. These substances are so named because they can be solubilized by solvents (ethanol, hexane, water, etc.). Extractive compounds comprise a large variation in the type and amount found in different species of SCB (Table 1). SCB extractives have been reported in a range of 3-14%, what depends on the species, growth condition and age. Among the compounds, phenolics are responsible for protection, avoiding insect, fungi and bacteria attack. The main types of compounds are phenolics, fats, fatty acids, resin, waxes and lignans. Since extractives are not linked to the lignocellulosic matrix they are removed previously of the biomass analysis such as chemical composition. Considering extractives removal, not for analytical purposes, their content is determined by extraction with ethanol and water in separate cycles, or with a mixture of cyclohexane/ethanol (1:1, v/v). The inorganic fraction of the biomass is named ash content. The ash is determined by burning biomass in mufla and quantifying it gravimetrically. The ash content in SCB varies on a short range around 2% (Canilha et al., 2011; Brienzo et al., 2009; Santos et al., 2011), however, a higher amount of ash has been reported (Table 1). SCB when in contact with soil in the field, normally shows a higher amount of ash, which comes from an excess of soil and dust (Szczerbowski et al., 2015). Samples withdrawn from the field should be washed with water at room temperature with the intent of removing these components that are not from the biomass composition. The biomass inorganic components are calcium, magnesium, silica, sodium, potassium, strontium, manganese and phosphorous (Szczerbowski et al., 2015). Considering the increasing of the uses of biomass to electricity generation through combustion, it is predicted that the amount of residual ash generated in such process will also increase. This ash will cause environmental concerns related to its properly storage and disposal. The challenge in to using this waste will be to applying it as feedstock to high-value products.

Table 1. Sugarcane bagasse chemical composition

	Chemical composition (%, dry mass)					Reference
	Cellulose	**Hemicellulose**	**Lignin**	**Extractives**	**Ash**	
	46.2	27.8	22.1	-	1.0	Santos et al., 2011
	45.0	31.8	20.3	-	2.9	Zhao et al., 2009
	35.0	35.8	16.1	-	-	Sasaki, 2003
	39.6	24.0	22.4	5.0	1.3	Benjamin et al., 2013
	35.1	30.3	19.6	9.9	1.6	Benjamin et al., 2013
	40.7	31.5	14.4	7.4	0.8	Benjamin et al., 2013
	42.4	25.2	19.6	-	1.6	Brienzo et al., 2009
	47.3	26.4	14.1	11.4	-	Brienzo et al., 2015
	43.1	25.1	20.8	4.9	-	Brienzo et al., 2015
	40.9	27.4	14.6	8.4	-	Brienzo et al., 2015
	40.5	29.0	30.6	-	2.3	Zend et al., 2014
	45.5	27.0	21.1	4.6	2.2	Rocha et al., 2011
	45.0	26.4	23.4	2.7	2.5	Rocha et al., 2012
	36.8	28.3	24.7	6.8	3.1	Rocha et al., 2015
	45.6	26.5	21.4	3.6	2.9	Rocha et al., 2015
	44.7	28.8	19.5	3.2	3.6	Rocha et al., 2015
	38.5	27.8	17.7	2.7	8.8	Guilherme et al., 2015
	45.0	25.8	19.1	9.1	1.0	Canilha et al., 2011
	36.1	26.9	26.2	5.3	2.1	Saad et al., 2008
	44.4	22.9	17.5	14.1	-	Maryana et al., 2014
	45.0	31.8	20.3	-	-	Zhao et al., 2009
Minimum	35.0	22.9	14.1	2.7	0.8	
Maximum	47.3	35.8	30.6	14.1	8.8	
Average	42.0	27.7	20.0	6.4	2.4	

Cellulose (glucan): represented as sum of anhydromonomers of glucose; Hemicellulose: represented as sum of anhydromonomers xylose, arabinose and acetyl group; Lignin: represented as sum of soluble and insoluble lignin; (-) not determined.

Hemicellulose Structural and Chemical Properties

Sugarcane bagasse hemicellulose is predominantly constituted of xylan, which is composed of xylose units in the backbone. SCB hemicellulose is defined as L-arabino-(4-O-methyl-D-glucurono)-D-xylan (Sun et al., 2004). Xylose accounts more than 70% in the xylan, in average, with a reported range of 43 to 93%. This polysaccharide shows branch groups of arabinose, acetyl groups and uronic acids (Table 2). Xylan backbone is made up of xylose linked by β-1,4 linkages with branch groups at the positions C2 and C3 with arabinosyl, uronic acids and acetyl groups (Sun et al., 2004). Arabinose is the branch group that appears in a higher amount. It is linked to xylan backbone by α-1,2 or α-1,3 linkage; galactose is linked by β-1,5 linkage to the xylan backbone. Xylan branch groups can be associated to other groups, aromatic feruloyl (from ferulic and diferulic acids) and p-coumaroyl can be attached to arabinose residues at the O-5 position (Saulnier et al., 1995). Other branch groups appear linked to xylan of SCB, which is reported to contain branch groups, what suggests that the xylan of this biomass is a heteropolysaccharide. SCB is a heterogeneous material, the fractions of the culm contain different amounts of xylan (Brienzo et al., 2014). Furthermore, it could be suggested that the xylan characteristic and properties could be different in the SCB fractions.

The branch groups of xylan of SCB depend on the variety type (Benjamin et al., 2014) and also on the condition of the extraction method (Brienzo et al., 2009). The contents of arabinose, galactose, glucose, rhaminose and uronic acids are quite different on literature reports (Table 2). The more branched chain correlates to solubility, which is important considering molecule reactivity. The branches also collaborate to chain characteristics, such as charge, which depends on the substituent groups. Some works showed a high content of glucose as part of the xylan; however, it could be probably from glucan and pectic polysaccharides (Geng et al., 2006). The use of water or a dilute alkaline medium before a more concentrated extraction probably removes the more branched xylan (Table 2). The use of peroxide in alkaline medium provokes the acetyl content removal from the xylan chain (Brienzo et al., 2009). The alkaline medium breaks alpha-ether linkages of lignin (of phenolic units), releasing the associated molecule that can be a hemicelluloses chain.

The alkaline medium also cleaves ester linkages, that include acetyl groups and ferulic and diferulic acids that link xylan chains to lignin. SCB with high acetyl content is positive for XOS release through pretreatments, such as autohydrolysis and steam explosion. These pretreatments, with no addition of catalyst, perform well in biomass with high acetyl content breaking xylan in XOS and xylose.

Hemicelluloses are naturally associated to cellulose and covalently linked to lignin in LCM. Xylan lignin residual content varies between 0.7 to 13% for extraction with hot water, alkaline and peroxide medium (Table 2). SCB pretreated with Soda-Anthraquinone has been further investigated to identify the structure formed of lignin-carbohydrate-complex (LCC) content. The LCC isolated were based on xylan-lignin complex and glucan-lignin-xylan (Njamela et al., 2013). The authors also showed that the ester linkages were broken during the hemicelluloses extraction. The isolation of xylan with alkali or peroxide in alkaline medium can show impurities related to lignin association. The association of lignin and hemicelluloses is done by several types of linkages that are resistant to alkaline or peroxide action. SCB has hydroxycinnamic acid (ferulic, coumaric and sinapic acids) involved in cross-linking xylan and lignin molecules (Rose, 2003). SCB xylan can be esterified to lignin. The hydroxyl groups of lignin units can be linked to carboxylic acid groups of uronic acids (Rose, 2003).

Native or extracted xylan contains molecules with a different number of xylose units, what means different degrees of polymerization (DP). The DP is defined as the number of xylose units in a xylan molecule. Xylan is reported as being made of anhydroxylose units linked by β-1,4 linkages with a DP up to 200. The xylan molecule is generally partially cleaved during the extraction process, with a consequent reduction in the DP. The DP values of xylan can vary according to the process method, its severity and the evaluation method (centrifugation, gel permeation, light scattering). The polysaccharide molecular mass (g/mol) is defined as the DP multiplied per the anhydroxylose molecular mass. The SCB xylan molecular mass has been reported in a wide range, including values between 6500 to 86000 g/mol (Table 2).

Table 2. Sugarcane bagasse hemicellulose chemical composition and extraction methods

Xylan composition (%, dry mass)								Molecular mass (g/mol)	Extraction yield (%)	Extraction condition	Reference
Xyl	Ara	Gal	Man	Rha	Glu	Uronic acids	Lignin				
50.4	4.4	-	-	-	3.9	4.4	10.0	21100	52.0	H_2O_2 in alkaline medium (2%, m/v), 20°C, 4h.	Brienzo et al., 2009
74.9	6.0	-	-	-	6.4	5.2	9.6	21100	60.8	H_2O_2 in alkaline medium (6%, m/v), 60°C, 16h.	Brienzo et al., 2009
78.0	5.8	-	-	-	5.4	6.5	10.4	22185	94.0	H_2O_2 in alkaline medium (4%, m/v), magnesium sulfate (0.25% m/m), 40°C, 10h.	Brienzo et al., 2009
93.2	5.6	0.2	0.4	0.2	0.5	2.3	6.1	40770	-	Succescive water, 1 and 3% NaOH, 50°C during 3h, precipitation with 15% ethanol.	Peng et al., 2009
91.0	6.4	1.3	0.3	0.1	1.1	2.1	4.5	86720	-	Succescive water, 1 and 3% NaOH, 50°C during 3h, precipitation with 30% ethanol.	Peng et al., 2009
87.0	7.8	1.1	nd	nd	3.3	1.7	2.2	77140	-	Succescive water, 1 and 3% NaOH, 50°C during 3h, precipitation with 60% ethanol.	Peng et al., 2009
82.0	9.3	0.7	1.0	3.0	4.1	4.8	3.1	35200	55.5	1 M NaOH, 20°C, 18h	Xu et al., 2006
81.0	9.8	0.5	1.1	4.9	2.6	5.8	2.7	35820	59.1	1 M NaOH, 30°C, 18h	Xu et al., 2006
78.0	11.6	0.3	1.4	6.5	2.2	5.6	2.4	37430	62.1	1 M NaOH, 40°C, 18h	Xu et al., 2006
43.1	25.1	2.7	4.0	0.7	9.8	2.1	13.3	6500	3.0	Hot water, 170°C, 15 min.	Banerje et al., 2014
69.0	12.1	1.4	1.1	0.1	9.7	1.7	6.3	22500	8.1	Hot water, 200°C, 15 min.	Banerje et al., 2014

Table 2. (Continued)

Xylan composition (%, dry mass)								Molecular mass (g/mol)	Extraction yield (%)	Extraction condition	Reference
Xyl	Ara	Gal	Man	Rha	Glu	Uronic acids	Lignin				
78.8	4.6	1.0	0.6	0.1	3.1	3.2	7.2	27900	14.9	Hot water extracted material 15 min, 170°C, and 3% H_2O_2, 0.25% magnesium sulfate, 40°C, 12h.	Banerje et al., 2014
83.4	5.7	0.9	0.4	0.2	3.0	3.4	3.9	12100	12.2	Hot water extracted material 15 min, 200°C, and 3% H_2O_2, 0.25% magnesium sulfate, 40°C, 12h.	Banerje et al., 2014
55.2	10.7	7.6	4.1	1.8	20.4	7.0	9.6	7380	4.1	Water extraction 55°C, 2h.	Sun et al., 2004
68.6	12.7	1.9	-	-	7.3	3.5	0.7	23340	9.6	Successive extractions of the 0.5 M NaOH treated bagasse with 0.5% H_2O_2, 55°C, 2h.	Sun et al., 2004
76.5	14.7	3.0	0.3	1.0	13.0	4.8	4.2	38890	9.6	Successive extractions of the 0.5 M NaOH treated bagasse with H_2O_2 (0.5; 1; 1.5; 2; 2.5 and 3%), 55°C, 2h.	Sun et al., 2004
81.3	11.9	1.7	0.5	0.5	4.0	3.5	5.1	45370	12.0	0.5 M NaOH at 55°C, 2 h from the water treated bagasse.	Sun et al., 2004
43.1	4.4	0.2	0.3	0.1	0.5	1.7	0.7	6500	3.0	Minimum	
93.2	25.1	7.6	4.1	6.5	20.4	7.0	13.3	86720	94.0	Maximum	
74.8	9.7	1.7	1.3	1.6	5.9	4.0	6.0	33026	32.6	Average	

(-) not detected or determined; Xyl: xylose; Ara: arabinose; Gal: galactose; Man: mannose; Rha: rhammnose; Glu: glucose.

XYLOOLIGOSACCHARIDES PREBIOTIC POTENTIAL AND HEALTH BENEFITS

Xylooligosaccharides are oligomers comprised of xylose units containing β-1,4 bonds. They are usually marketed as a white powder containing 2 to 12 xylose units (Figure 1), although up to 20 units molecules are still considered XOS (Mäkeläinen et al., 2010a; Moure et al., 2006; Vázquez et al., 2000; Kabel et al., 2002). The composition and stability of XOS are related to the type of oligosaccharide, sugar residues, bonds, ring formation and anomeric configurations and the extraction process (Kabel et al., 2002; Carvalho et al., 2013). The XOS present some advantages compared to fructooligosaccharides and inulin, such as resistance to a broad range of pH (2.5-8.0) and stability at high temperatures, above 100°C. The lack of the enzyme that cleavages β bonds in humans and the stability at acidic pH allow the XOS to reach the intestines intact, serving as a substrate for beneficial bacteria colonizing the intestinal tract (Bielecka et al., 2002). This feature allows the XOS to be classified as non-digestible oligosaccharides (NDOS), which are soluble fibers. The NDOS are water soluble and generally present sweetness (Crittenden and Playne, 1996; Manning and Gibson, 2004).

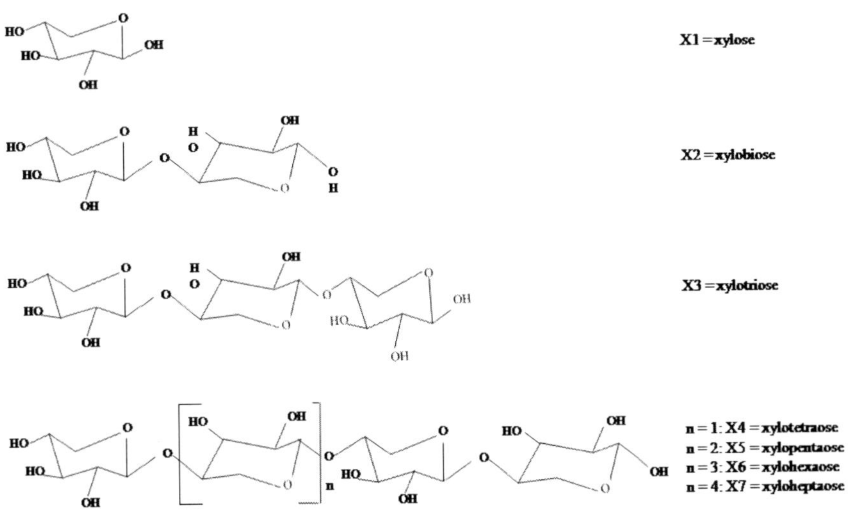

Figure 1. Schematic structure of xylose and xylooligosaccharides.

The gastrointestinal tract is heavily colonized by various microorganisms, including bacteria with pathogenic potential, as well as beneficial bacteria. Due to their beneficial effects, and especially due to the ability to selectively stimulate the growth of *Bifidobacterium* and *Lactobacillus*, XOS are considered prebiotic compounds (Moure et al., 2006, Mussatto e Mancilha, 2007). Many prebiotics are commercialized worldwide, however this requires that the prebiotic be resistant to gastric juice and to digestive enzymes present in the gastrointestinal tract; able to be fermented by the intestinal flora; and selectively stimulate the growth or activity of intestinal bacteria beneficial to health (Gibson et al., 2004).

Nowadays, a great interest in a healthier lifestyle has risen, once the development of foods that bring health benefits has become the main objective of the food industries. XOS can be used for the production of xylitol (sweeteners), that act in the prevention of tooth decay and can be added to food for consumption by people with diabetes or in need of low calorie diets (Rivas et al., 2002). Apart from the prebiotic effect, studies indicate improvements in bowel function, an increase in the absorption of nutrients, fortification of the immune system, prevention of cardiovascular diseases, reduction of colon cancer risk and the triggering of allergenic, anti-inflammatory and anticariogenic activity (Grootaert et al., 2007; Wang et al., 2009; Aachary and Prapulla, 2009).

Another aspect of XOS usage is related to animal feeding. The use of antibiotics in livestock aims to fight infections and help animals gain weight. However, the excessive use of antibiotics enables resistance and production of virulent strains (Carvalho et al., 2013). In general, the use of prebiotics as an additive in animal feeding has attracted attention due to their effects on animal health. The addition of prebiotics on chicken ration is gaining prominence as organic additives since it shows positive effects similar to antibiotics, but without causing bacterial resistance (Flemming and Freitas, 2005). Studies *in vitro* demonstrated the positive effects of XOS digestibility in the intestine of ruminants, since they stimulated the growth of probiotic strains (Samanta et al., 2012). Beneficial changes in the intestinal microbiota reduce the population of some potentially pathogenic strains of the genus *Escherichia*, *Salmonella*, *Enterobacteriaceae* and *Streptococcus*, bacterial agents of intestinal disorders and toxins release (Holzapfel, 1998). Prebiotics are also studied because of health benefits they provide during pregnancy. Studies *in vivo* with rats fed with a combination of FOS and XOS showed that the supplementation with prebiotics during gestation can protect the fetus, developing brain against oxidative stress and also restoring enzymic antioxidants and mitochondrial function (Krishna et al., 2015).

The studies of XOS biological properties and prebiotic effects have been done *in vitro* and *in vivo* for humans and animals. The assay *in vitro* considers microorganisms common in the human and animal guts, evaluating the potential of stimulating the beneficial and inhibiting the pathogenic microorganisms. Recent tests *in vitro* showed that XOS derived from corn cobs improve the growth of *Lactobacillus plantarum* S2, and the combination of both prebiotic and probiotic causes inhibiting effects, decreasing the viability of *Enterococcus*, *Enterobacter*, and *Clostridia* spp. This combination also showed an antioxidant activity better than that of XOS or *L plantarum* alone (Yu et al., 2015). Several species of *Bifidobacterium*, *Lactobacillus brevis* and some *Bacteroides* are able to grow in culture media containing XOS with 2-5 units of xylose as a sole carbon source. Moreover, XOS can inhibit *Escherichia coli*, *Enterococcus spp.*, *Clostridium perfringens* and *Clostridium difficile* (Crittenden et al., 2002). Studies on pure cultures and on simulated colon models showed that XOS selectively stimulate the growth of *Bifidobacterium lactis* and some species of *Lactobacillus* (Mäkeläinen et al., 2010a, b). XOS are also able to inhibit the adhesion of *Listeria monocystogenes* on the intestinal epithelium (Erbesbach et al., 2011). Therefore, the quality of the produced XOS is important when considering the prebiotic effect.

Commercial XOS (Wako Chemicals) and the one produced from *Miscanthus giganteus,* fermented *in vitro* with human fecal microbiota, demonstrated an increase in bacteria of genus *Bifidobacterium*, *Lactobacillus* and *Escherichia* (Chen et al., 2016a). Another work using XOS produced from *Miscanthus giganteus* demonstrated that *Bifidobacterium adolescentis* and *Bifidobacterium catenulatum* grew in XOS, and acetic and lactic acids were the products of this fermentation (Chen et al., 2016a). Commercial XOS (Xylooligo 95P; Suntory, Osaka, Japan) produced by autohydrolysis, containing 83% of xylobiose and xylotriose were used in culture medium as a carbon source. These XOS were responsible for higher growth of *Bifidobacterium* strains (*B. adolescentis* and *B. longum*) than the culture with medium formulated with another XOS with higher DP and containing only 24–41% of xylobiose and xylotriose. Furthermore, XOS with DP 5 and 6 (longer chain lengths) reduced the degree of consumption of these oligosaccharides by the bacteria (Moura et al. 2007).

Xylooligosaccharides dosage and DP are important issues that have been investigated by several researchers. The key factor for the use of XOS in human and animal feed is to determine the required daily dosage and the best range of DP to maximize its effects. Tests carried out *in vitro* and *in vivo* using oral administration of XOS (DP 2 and 3, with xylose) of 5g/day for 3 weeks indicated that XOS reached the intestine intact and were able to stimulate the growth of bacteria of the *Bifidobacterium* genus, particularly *B. adolescentis*. Furthermore, XOS inhibited the growth of *Clostridium* and *E. coli* (Okazaki, 1990). Studies *in vivo* showed that a low dose of commercial XOS (Shandong Longlive Bio-Technology Co., Ltd., China.) is enough to increase *Bifidobacterium* in the human guts. Only 1.4 g/day increased the counts of *Bifidobacterium* compared to a placebo group, and 2.8 g/day showed an even greater increase, without any gastrointestinal side effects (Finegold et al., 2014). This study did not observe changes in the *Lactobacillus*, *Clostridium* and *Enterobacteriaceae* counts, or in pH and short chain fatty acids related to XOS consumption. The maximum tolerant dose of XOS consumption for humans was determined as 12 g a day. With this amount of daily XOS, side effects such as diarrhea and constipation were observed to be similar to those subjects who didn't take any XOS (Xiao et al., 2012).

HEMICELLULOSE EXTRACTION METHODS

The hemicellulose extraction method depends on the purpose and application. The selection of the extraction method should be based on the desirable final characteristics of the hemicellulose. Among these important characteristics for application are the degree of polymerization, degree of branching/substitution, impurity such as residual lignin content, solubility and reactivity. Considering hemicellulose recovery based on polysaccharide or monosaccharide form, alkaline or acid concepts can be successfully applied. Hemicellulose extraction with acid medium and process derived from it lead to a final product based on monosaccharide and oligosaccharides. On the other hand, hemicelluloses extraction with alkaline medium and process derived from it lead to a final product based on polysaccharides with a high degree of polymerization. Considering XOS production, both methods could be applied. Acid medium results in low XOS yield high xylose yield, with by-products formation. Alkaline extracted hemicelluloses require a second step for XOS production, which could be an acid or an enzymatic hydrolysis. Despite the many possibilities for XOS production, it is still not economic and technically feasible for industrial scale (Carvalho et al., 2013).

Pretreatment is necessary in order to remove/modify lignin in LCM and to obtain sugars from cellulose or hemicellulose. This process is then followed by enzymatic hydrolysis to break the polysaccharides and consequently release the oligomers or monomers (Mosier et al., 2005; Balat et al., 2008). The pretreatment may act by altering structural factors or chemical composition of the biomass, besides increasing the surface area, facilitating the accessibility of the constituent LCM (Brienzo et al., 2015; Mosier et al., 2005). The chemical inhibitors may be present in the biomass or produced during the pretreatment, generally in severe conditions pretreatments. The sugars degradation forms organic acids that can be regarded as enzymatic hydrolysis inhibitors (Carvalho et al., 2015). The toxic compound 5-hydroxymethylfurfural (HMF) is derived from the degradation of hexoses, as furfural comes from pentoses. The breaking of furfural and HMF releases formic acid, in addition, HMF degradation produces levulinic acid, and phenolic compounds are formed from the lignin (Rasmussen et al., 2014). After the pretreatment step, alcohols and ketones are used for the recovery of solubilized hemicellulose (Akpinar et al., 2009; Vázquez et al., 2005). However, XOS obtained through certain methods need a more complex purification step due to the by-products and xylose present in the hydrolysate.

Acid and Hydrothermal Pretreatment

The acid hydrolysis is often performed with diluted acid to prevent corrosion of equipments and high energy demand for the acid recovery. The pretreatment with diluted acid results in less formation of furfural and hydroxymethylfurfural (HMF) (Gómez et al., 1985). The most commonly used acids in this process are sulfuric, hydrochloric, acetic and phosphoric acids. In this pretreatment, hemicellulose is solubilized, containing predominantly pentose and a fraction of an insoluble part, that contains mainly cellulose and lignin. The acid concentration should be low, in order to avoid hemicellulose hydrolysis into monomers, such as xylose, and consequently the release of compounds from its degradation, like furfural.

Acid hydrolysis is able to cleave the glucosidic linkage, fractionating the polysaccharides to oligosaccharides or monosaccharide (Figure 1). The acid catalysis occurs in three steps: a rapid protonation of the oxygen atom of the ether linkage (C-O-C); the transference of the positive charge to the C1, forming a carbocation ion and simultaneous split of the ether linkage; and a fast addition of water to the carbocation ion. The β-1,4 linkage in xylan is susceptible to acid attack, provoking chain breakdown, which depends on the acid type and concentration, and reaction condition, such as temperature and time. The acid-catalyzed xylan chain hydrolysis can happen through random attack and depolymerization. The random cleavage breaks the chain, decreasing the degree of polymerization and releasing oligosaccharides. The depolymerization depends on the chain DP and releases monosaccharides. These two types of acid attack occur together and need to be controlled to maximize the desired product: xylose or XOS.

Hydrothermal pretreatment is catalyzed by steam or water hydronium (H_3O^+), process known as hydrothermolysis, autohydrolysis or hotwater (Figure 2). Autohydrolysis can be carried out at temperatures between 150 and 220°C. This process has the advantage of not using harsh chemicals to extract hemicellulose, avoiding corrosion of the equipment. This process is mainly applied in agro-industrial waste, in which hemicelluloses are mainly xylan (Overend and Chornet, 1989). The amount and concentration of solubilized products distinguish autohydrolysis in hydrothermal pretreatment and steam explosion. Steam explosion is performed by the sudden decompression of a pressurized system containing high pressure water steam and biomass (Hendriks and Zeeman, 2009). The amount of products from the degradation of sugars is smaller in the hydrothermal pretreatment than in steam explosion (Taherzadeh and Karimi, 2008).

The autohydrolysis is a process with no catalyst-added, and the process occurs under acidic condition, due to acetic acid released by cleavage of the acetyl groups substituent in the xylan chain. The acetic acid released creates a slight acid medium with pH around 4, and its release is a temperature-based reaction of ester linkages, which are unstable at high temperature (Fengel and Wengels, 1984). The severity of the autohydrolysis process is determined by the temperature and the reaction time. These two variables are directly related to the acetil groups originated from hemicellulose (Garrote, 1999). Considering that the acetyl groups on hemicellulose are responsible for acidifying the medium, it could be suggested that the amount of acetyl on the biomass will lead to different responses on XOS production (Nabarlatz et al., 2007). Considering the autohydrolysis process for XOS production, the feedstock material characteristics are important for the final yield. The autohydrolysis is successful for biomass with high acetyl content. On the other hand, acid catalyzed could be desired for biomass with low acetyl content. The water dissociation to form hydronium has been suggested, however the main factor involved in this ion formation is the acetic and uronic acids released from the biomass (Nabarlatz et al., 2004).

The hydrothermal pretreatment, catalyzed or not, releases xylan from the biomass in the form of soluble XOS, which are in sequence breakdown on xylose. Depending on the reaction condition (severity, i.e., temperature, reaction time and pH), the xylose can be degraded to furfural, which can be converted into other degradation and condensation products (Rasmussen et al., 2014). The xylan solubilization and hydrolysis to XOS, xylose and subsequent conversion to degradation products (by-products) could occur in a sequence of consecutive reactions (Figure 3). However, the LCM is heterogeneous and the interaction between xylan and the other macromolecules can be different, considering different material fractions and cell wall organization. The SCB has fractions with different xylan contents (Brienzo et al., 2014), and it is possible that its recalcitrance is different, implying fast and slow reaction for XOS and xylose releasing.

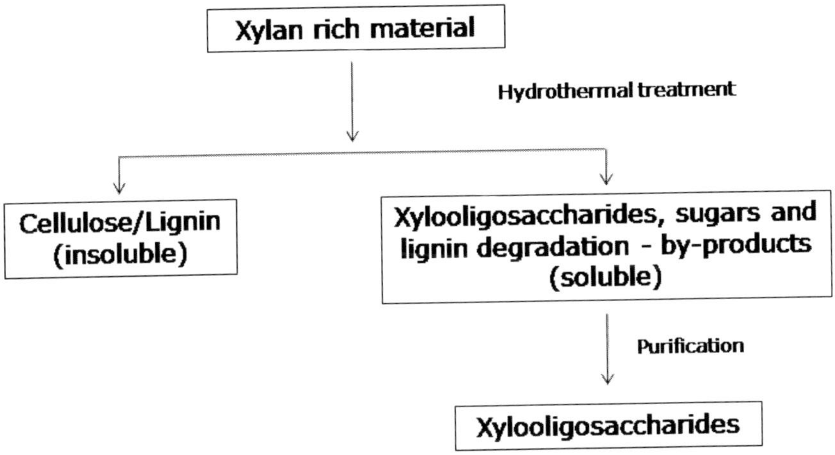

Figure 2. General scheme for XOS production by hydrothermal treatment.

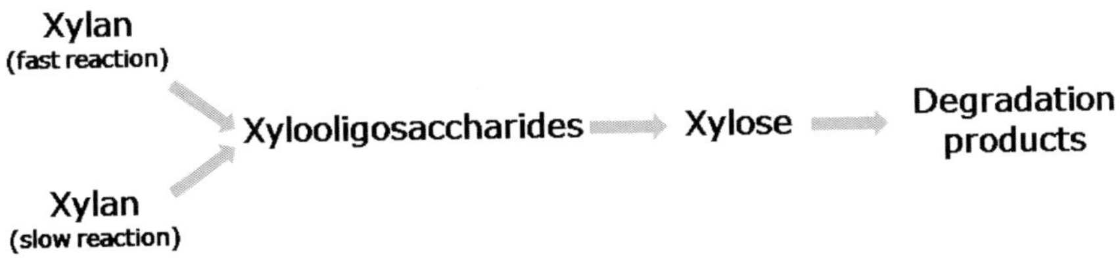

Figure 3. Xylan acid hydrolysis and possible steps. A complex scheme should be developed considering the recalcitrance and material heterogeneity. The xylan can be at different degrees: soluble, acetyl content, branches, accessible and lignin protected.

Hydrothermal pretreatment has been described as efficient for XOS production because of high yield (Otieno and Ahring, 2012; Chen et al., 2014, Moniz et al., 2014). On the other hand, there is a production of undesirable compounds, such as monosaccharides (xylose) and the degradation products of sugars and lignin. The releasing of by-products and monosaccharides requires a purification step, which accounts on process cost, considering industrial scale. Carbon adsorption could recover 45% of XOS, while serial ion exchange resin treatments could recover 91% of XOS (Chen et al., 2016b). A hydrothermal pretreatment acid catalyzed (0.1% of H_2SO_4, pre-soaked at 60°C/12 h, followed 145°C/1 h) was efficient for several biomass, releasing 92% of XOS from SCB (Otieno and Ahring, 2012). These XOS showed DP ranging from 2 to 25, with higher presence of DP 2 to 5. Hydrothermal pretreatments are dependent on the severity factor, which should be adjusted to maximize XOS release. A process with low severity is characterized by low XOS, however, with low by-products release. High severity also is characterized by low XOS production, because of its breakdown, but high xylose and by-products release (Moniz et al., 2014).

Alkaline Hemicelluloses Extraction

The alkaline extraction breaks down the cell wall, disrupting the macromolecule components, so making xylan and lignin be solubilized. The reaction normally uses a strong solution of NaOH, KOH, $Ca(OH)_2$, NH_3 and also oxidative agents, i.e., H_2O_2 in alkaline medium. The effectiveness of the xylan solubilization depends on the lignin content in biomass. The alkaline action is assigned to a saponification of the ester linkage, separating the hemicellulose from the lignin, and releases uronic and acetic acid. In addition, the pretreatment of the biomass with a solution of NaOH causes a swelling of the cellulose structure and accordingly increases the surface area (Weil et al., 1994). This increased structure facilitates the intake of water, which acts disrupting the interaction between lignin and carbohydrates. The result is a decrease in crystallinity, and the rupture of the lignin that is soluble in these conditions (Weil et al., 1994; Balat et al., 2008). Compared to acid pretreatment methods, the alkaline methods have the advantage of removing the lignin without degrading the other components, usually due to milder process conditions, which can be carried out even at room temperature (Balat et al., 2008).

The H_2O_2 reacts with the aromatic and aliphatic structures of lignin, and it is widely used in pulp bleaching processes. The action of this oxidant agent can generate degradation products due to the formation of oxygen radicals. The influence of pH has been extensively studied and the literature shows the value of 11.6 (Yang et al., 2002). In alkaline medium, the hydrogen peroxide decomposes in hydroperoxide anion (HOO^-), the main agent in the degradation of lignin compounds. On the other hand, when the peroxide is used as the delignification agent in alkaline extraction, the decomposition occurs due to the formation of hydroxyl and oxygen, which begins lignin breakdown. Although the decomposition of peroxide is necessary for delignification to occur, the rate of decomposition must be reduced to avoid excessive formation of free radicals, which can lead to degradation of polysaccharides. The peroxide pretreatment can be favored by the removal of heavy metals with chelating agents, and adding later magnesium ions. The metals catalyze the decomposition of peroxide on hydroxyl radicals, provoking polysaccharide depolymerization and decrease the final yield; the magnesium ions stabilize the peroxide avoiding degradation (Thakore et al., 2005; Brienzo et al., 2009).

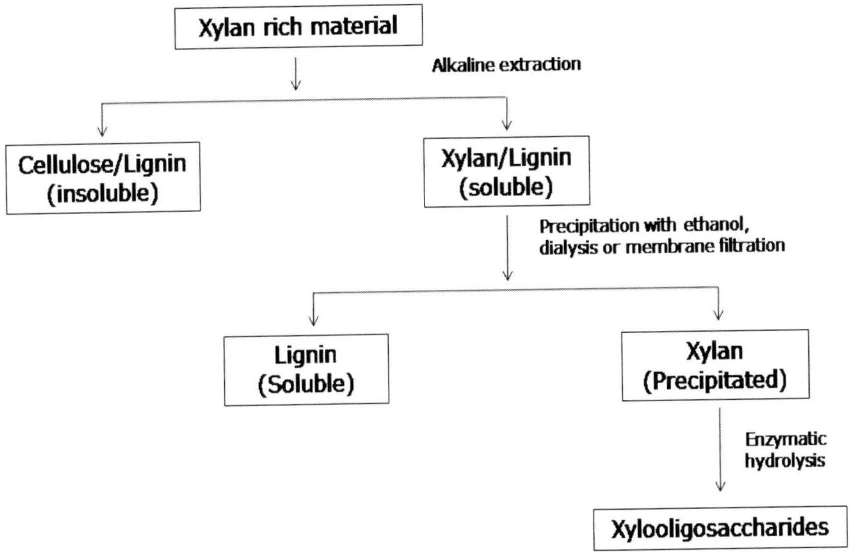

Figure 4. General scheme for XOS production by two-stage: hemicelluloses extraction and enzymatic hydrolysis.

Depending on the process condition, a high yield of hemicellulose can be extracted (Table 2). Once hemicellulose is isolated, the enzymatic hydrolysis act efficiently release XOS. This process is characterized in two-stage: an alkaline extraction of the hemicelluloses followed by an enzymatic hydrolysis (Figure 4).

Enzymatic Hydrolysis

Refering to lignocellulosic materials (LCM), enzymatic hydrolysis is a widely studied process, since this type of hydrolysis presents the specificity of the reaction and absence of secondary reactions, which would have decreased the yields. Another important characteristic of this hydrolysis is the lack of by-products formation, since the reactions happen mildly and there is no need of high pressures or temperatures, nor of the use of corrosive materials (Figure 4). However, the conversion of LCM into sugars through an economic process is still a challenge. The demand for hydrolytic enzymes is growing faster than ever and has become a driving force for researches on xylanases and cellulases. However, the production costs and low yields are challenges for industrial applications. The incorporation of economical sources (LCM such as SCB, wheat straw, corn cobs) in the culture medium for the growth of microorganisms and enzyme production may help decrease the production cost of the enzymes and special microorganisms can also be used as a biological pretreatment of LCM. Consequently, several processes are being developed using agro-industrial residues for the production of these enzymes (Lakshmi et al. 2009). Moreover, the bioconversion of these substrates may help the reduction of the environmental impact caused by the accumulation of waste (Camassola and Dillon 2009).

Although decades have already been dedicated to LCM conversion, the substrate characteristics that give more hydrolysis efficiency are still not completely understood. The enzymatic hydrolysis requires direct contact between the enzyme and the substrate. The enzyme must overcome physical barriers to adsorb on the surface of the substrate and finally catalyze the hydrolysis. LCM is naturally resistant to enzyme or microorganism action due to its highly organized cellulose-hemicellulose-lignin structure in the plant cell wall (Sant'Anna et al., 2014). Substrates with high lignin content and highly crystalline have low enzyme effectiveness. The factors that influence the performance of enzymatic hydrolysis process depend on the: (1) substrate structure, which is influenced by the substrate/material characteristics; (2)

interactions of the enzymes with the substrate, that depend on the nature and source of the enzyme complex. In addition to these factors, the enzymatic hydrolysis is influenced by the heterogeneity of the LCM. A better enzyme action could be reached with a previous isolation of the polysaccharide. Enzyme applied to XOS production requires hemicellulose extraction for further hydrolysis, characterizing a two-stage process (Brienzo et al., 2010).

The hydrolysis of hemicellulose takes place through the action of endo-enzymes that act internally in the main chain and exo-enzymes, responsible for the cleavage that produces monosaccharides and oligosaccharides (Kalogeris et al., 2001). The endo-β-1, 4-D-xylanase (EC 3.2.1.8) acts on the main chain of xylan and generates XOS with low DP, which are substrates for exo-β-1,4-xylanase (EC 3.2. 1:37), that act on the non-reducing terminal D-xylose (Beg et al., 2001, Lakshmi et al. 2009). Hemicellulose normally is a heteropolymer with several branching units linked to the backbone. For the complete hydrolysis of the hemicelluloses different enzymes are necessary in synergistic action. Enzymes that act on the branches (dibranching or accessory enzymes) are arabinofuranosidase (EC 3.2.1.55), able to remove arabinose. α-glucuronidase (EC 3.2.1.131) removes glucuronic and acetyl xylan (Lagaert et al., 2014). Acid esterase (EC 3.1.1.72) removes acetyl groups (Juhász et al., 2005). According to enzymes action, for XOS production, it is desired an enzymatic pool with no exo-β-1,4-xylanase activity. The presence of this last enzyme will provide a release of xylose, decreasing the yield of XOS.

The presence of dibranching enzymes can lead to generate XOS with no branch units. Branched XOS can be released with enzymatic hydrolysis free of accessory enzymes (Puchart and Biely, 2008). In fact, the branch groups influence on the xylan hydrolysis rate, that is strongly dependent on the amount of branch units (Li et al., 2007). The endo-xylanases act differently on the xylan backbone, once they are influenced by the branch groups. Endo-xylanase from Glycoside Hydrolase (GH) Family 10 produces XOS with lower DP than GH family 11 (Maslen et al., 2007). The presence of branching can be a barrier, depending on the enzyme restriction due to the mode of action. Xylanase from *Aspergillus niger* hydrolyzes better the unbranched xylan, while xylanase from *Trichoderma longibrachiatum* can hydrolyze both types of xylan (Akpinar et al., 2009). Xylanases purified from *A. niger* could hydrolyze xylotriose, but is not able to hydrolyze arabinoxilotriose (Takenishi and Tsujisaka, 1975). The presence of substituents, which are typically located in the non-reducing terminal of XOS, protects the glycosidic bond of the main chain (Contat and Joseleau, 1981). The presence of branch groups can result in diverse biological properties. The branched XOS has shown a higher selectivity by bifidobacteria compared to none branched (Van Laere et al., 1997). Currently, there is a lack of information and studies dedicated to elucidate the mechanism of the XOS and branched-XOS on the biological properties. XOS can be responsible for several biological properties different from the prebiotic effects: immunomodulatory, anti-cancerous, anti-microbial and growth regulator. Some of the properties are suggested to be related to XOS branch groups, for example, antioxidant activity is due to phenolic substituents as a branch; anti-allergy and anti-inflammatory properties are attributed to uronic substituents (Yuan et al., 2004; Izumi et al., 2004).

The optimization of the conditions of hemicellulose hydrolysis into XOS by xylanases is also another important step to improve the XOS technology, since a higher yield is necessary to obtain an economic bioprocess (Brienzo et al. 2009 and 2010; Akpinar et al. 2009; Carvalho et al. 2013; Carvalho et al. 2015). In addition, the presence of a low amount of β-xylosidase (β-1,4-xylan xylohydrolase, EC 3.2.1.37) in xylanase extracts is essential, since this enzyme converts xylobiose and other XOS into xylose (Vázquez et al. 2000). Some microbiological and biochemical aspects of xylanase hydrolysis can influence the XOS production. The improvement of the enzymatic reaction can be performed to obtain higher XOS yield and also reach a concentration by optimization of the dosage of xylanase, reaction time, temperature and substrate concentration. The hemicellulose studies with various extraction methods and followed hydrolysis revealed a maximum XOS concentration of 7.5 g/L using 2% of hemicellulose obtained from bagasse, 120-500 U/g xylanase during 48-72 h (Carvalho et al, 2015); 5.7 g/L of XOS after 96 h, 60 U/g using 2.6% of hemicellulose (Brienzo et al. 2010) and 1.7 g/L after 8 h (Jayapal et al. 2013). The XOS production with other substrates reached maximum production with corncob, 5.8 g/L after 24 h (Aachary and Prapulha, 2009) and 6.7 g/L after 8 h (Chapla et al. 2012).

The enzymatic hydrolysis process was suggested as the best option for XOS production, considering the release of a low amount of monosaccharides and undesirable products (by-products) (Aachary and Prapulla, 2009). The xylanases from *A. fumigatus* M51 reached a high level of XOS production (37.6%) after 48–72 h using hemicellulose extracted from SCB. This yield represents 68.8 kg of XOS with prebiotic activity (xylobiose and xylotriose) per metric ton of SCB (Carvalho et al. 2015). However, the best method for XOS production is not defined, since the enzymatic process is lower and probably less economic than the combination of acid and hydrothermal method. On the other hand, the enzymatic method resulted in the production of XOS with a degree of polymerization of 2–3, without by-products, while hydrothermal pretreatment produced other XOS with higher DP without prebiotic effect, and some by-products, such as furfural and hydroxymethylfurfural (Otieno and Ahring, 2012; Chen et al., 2014, Muniz et al., 2014).

CONCLUSION

The worldwide current concern about the environmental protection leads to the biotechnological use of LCM. SCB is one of the most important wastes generated from the agro-industrial process. Among several possibilities of the SCB use, the hemicellulose extraction to be used as feedstock for high-value molecule production has gained attention. Hemicellulose has been studied for several applications in the field of energy, food and pharmaceutical products. Among them, the production of XOS has received special attention because of its prebiotic effect and for being considered as an ingredient in functional foods. They are favorable for the health of consumers as they are not cariogenic, have low calorific value and stimulate the growth of beneficial bacteria in the colon, as well as several biological properties can be provided. The hemicellulose extraction is not a simple task and still has challenges to reach the industrial application. The available process for hemicellulose extraction has to solve problems such as high amount of chemicals used, low extraction yield, sugars degradation and residual lignin. Further, the XOS generation through hydrothermal and enzymatic hydrolysis processes needs to be improved to overcome the material recalcitrance. The hydrothermal pretreatment has the challenge of avoiding sugars degradation products, and the control of hemicellulose depolymerization, avoiding xylose release or no prebiotic high DP XOS production. Despite the improvement in the technology of enzymatic hydrolysis, with the predominant production of prebiotic XOS and no toxic compounds generated, this process remains expensive, and it depends on the hemicellulose pre-extraction. The growing importance, biological properties and demand for XOS are opportunities for process development, leading to efficient hemicellulose extraction and enzymatic hydrolysis processes, or a safe hydrothermal process for XOS production.

REFERENCES

Aachary, A. A. and Prapulla, S. G. (2009). Value addition to corncob: Production and characterization of xylo-oligosaccharides from alkali pretreated lignin-saccharide complex using *Aspergillus oryzae* MTCC 5154. *Bioresour Technol.*, *100*, 991-995.

Akpinar, O., Erdogan, K. and Bostanci, S. (2009). Enzymatic production of xylooligosaccharide from selected agricultural wastes. *Food Bioprod Process.*, *87*, 145-151.

Balat, M., Balat, H. and Öz, C. (2008). Process in bioethanol processing. *Prog Energy Combust Sci.*, *34*, 551–573.

Banerjee, P. N., Pranovich, A., Dax, D. and Willför, S. (2014). Non-cellulosic heteropolysaccharides from sugarcane bagasse – Sequential extraction with pressurized hot water and alkaline peroxide at different temperatures. *Bioresour Technol.*, *155*, 446-450.

Beg, Q. K., Kapoor, M., Mahajan, L. and Hoondal, G. S. (2001). Microbial xylanase and their Applications: a review. *Appl Microbiol Biotechnol.*, *56*, 326-338.

Benjamin, Y., Cheng, H. and Görgens, J. F. (2013). Evaluation of bagasse from different varieties of sugarcane by dilute acid pretreatment and enzymatic hydrolysis. *Ind Crops Prod.*, *51*, 7–18.

Benjamin, Y., Cheng, H. and Görgens, J. F. (2014). Optimization of dilute sulfuric acid pretreat-ment to maximize combined sugar yield from sugarcane bagasse for ethanol production. *Appl Biochem Biotechnol.*, *172*, 610–30.

Bielecka, M., Biedrzycka, E., Majkowska, A., Juśkiewicz, J. and Wróblewska, M. (2002). Effect of non-digestible oligosaccharides on gut micro ecosystem in rats. *Food Res Int.*, *35*, 139–144.

Brienzo, M., Carvalho, W. and Milagres, A. M. F. (2010). Xylooligosaccharides production from alkali-pretreated sugarcane bagasse using xylanases from Thermoascus aurantiacus. *Appl Biochem Biotech.*, *162*, 1195–205.

Brienzo, M., Ferreira, S., Vicentim, M. P., Souza, W. and Sant'Anna, C. (2014). Comparison study on the biomass recalcitrance of different tissue fractions of sugarcane culm. *Bioenerg Res.*, *7*, 1454-1465.

Brienzo, M., Siqueira, A. F. and Milagres, A. M. F. (2009). Search for optimum conditions of sugarcane bagasse hemicellulose extraction. *Biochem Eng J.*, *46*, 199-204.

Brienzo, M., Tyhoda, L., Benjamin, Y. and Görgens, J. (2015). Relationship between physicochemical properties and enzymatic hydrolysis of sugarcane bagasse varieties for bioethanol production. *N Biotechnol.*, *32*, 253-262.

Camassola, M. and Dillon, A. J. P. (2009). Biological pretreatment of sugar cane bagasse for the production of cellulases and xylanases by Penicillium echinulatum. *Ind Crop Prods*, *29*, 642–647.

Canilha, L., Santos, V. T. O. and Rocha, G. J. M. (2011). A study on the pretreatment of a sugarcane bagasse sample with dilute sulfuric acid. *J Ind Microbiol Biot.*, *38*, 1467-1475.

Carpita, N. C. and Gibeaut, D. M. (1993). Structural models of primary cell walls in flowering plants: consistency of molecular structure with the physical properties of the cell wall during growth. *Plant J.*, *3*, 1-30.

Carvalheiro, F., Esteves, M. P., Parajó, J. C., Pereira, H. and Girio, F. M. (2004). Production of oligosaccharides by autohydrolysis of brewery's spent grain. *Bioresour Technol.*, *91*, 93–100.

Carvalho, A. F. A, Neto-Oliva, P., Silva, D. F. and Pastore, G. M. (2013). Xylo-oligosaccharides from lignocellulosic materials: Chemical structure, health benefits and production by chemical and enzymatic hydrolysis. *Food Res Int.*, *51*, 75-85.

Carvalho, A. F. A., Oliva-Neto, P., Almeida, P. Z., Silva, J. B., Escaramboni, B. and Pastore, G. M. (2015). Screening of xylanolytic *Aspergillus fumigatus* for prebiotic xylooligosaccharide production using Bagasse. *Food Technol Biotech.*, *53*, 428-435.

Chapla, D., Pandit, P., Shah, A., (2012), Production of xylooligosaccharides from corncob xylan by fungal xylanase and their utilization by probiotics. *Bioresour Technol.* *115*, 215-21.

Chen, M. H., Swanson, K. S., Fahey, Jr. G. C., Dien, B. S., Beloshapka, A. N., Bauer, L. L., Rausch, k. D., Tumbleson, M. E. and Singh, V. S. (2016a). *In vitro* fermentation of xylooligosaccharides produced from *Miscanthus* × *giganteus* by human fecal microbiota. *J Agric Food Chem.*, *64*, 262–267.

Chen, M. H., Bowman, M. J., Cotta, M. A., Dien, B. S., Iten, L. B., Whitehead, T. R., Rausch, K. D., Tumbleson, M. E. and Singh, V. (2016b). *Miscanthus* × *giganteus* xylooligosaccharides: Purification and fermentation. *Carbohydr Polym.*, *140*, 96–103.

Chen, M. H., Bowman, M. J., Dien, B. S., Rausch, K. D., Tumbleson, M. E. and Singh, V. (2014). Autohydrolysis of *Miscanthus* × *giganteus* for the production of xylooligosaccharides (XOS): Kinetics, characterization and recovery. *Bioresour Technol.*, *155*, 359–65.

Chiang, V. L. and Funaoka, M. (1990). The difference between guayacyl and guaiacyl-syringyl lignins in their response to kraft delignification. *Holzforschung.*, *44*, 309-313.

Contat, J. and Joseleau, J. P. (1981). Mode of action a xylanase and its significance for the structural investigation of the branched L-arabino-D-glucurono-D-xylan from redwood (*Sequoia sempervirens*). *Carbohydr Res.*, *95*, 101-112.

Crittenden, R., Karppinen, S., Ojanen, S., Tenkanen, M., Fagerström, R., Mättö, J., Saarela, M., Mattila-Sandholm, T. and Poutanen K. (2002). *In vitro* fermentation of cereal dietary fibre carbohydrates by probiotic and intestinal bacteria. *J Sci Food Agric*, *82*, 781-789.

Crittenden, R. G. and Playne, M. J. (1996). Production, properties and applications of food-grade oligosaccharides. *Trends Food Sci Tech.*, *71*, 353-361.

Fengel, D. and Wegener, G. (1984). Wood: Chemistry, Ultrastructure reactions. New York: W. De Gruyter., 613.

Finegold, S. M., Li, Z., Summanen, P. H., Downes, J., Thames, G., Corbett, K., Dowd, S., Krak, M. and Heber, D. (2014). Xylooligosaccharide increases bifidobacteria but not lactobacilli in human gut microbiota. *Food Funct.*, *5*, 436-445.

Flemming, J. S. and Freitas, R. J. S. (2005). Avaliação do efeito de prebióticos (MOS), probióticos (*Bacillus licheniformis* e *Bacillus subtilis*) e promotor de crescimento na alimentação de frangos de corte. *Arch Vet Sci.*, *10*, 41-47.

Garrote, G., Domínguez, H. and Parajó, J. C. (1999). Mild autohydrolysis: an environmentally friendly technology for xylooligosaccharide production from wood. *J Chem Technol Biotechnol.*, *74*, 1101-1109.

Geng, Z. C., Sun, J. X., Liang, S. F., Zhang, F. Y., Zhang, Y. Y., Xu, F. and Sun, R. C. (2006). Characterization of water- and alkali-soluble hemicellulosic polymers from sugarcane bagasse. *Int J Polym Anal Chem.*, *11*, 209-226.

Gibson, G. R., Beatty, E. R., Wang, X. and Cummings, J. H. (1995). Selective stimulation of bifidobacteria in the human colon by oligofructose and inulin. *Gastroenterology.*, *108*, 975-982.

Gómez, R. J. H. Sacarificacao da hemicelulose do bagaco de cana-de-acucar e sua fermentacao por pachysolen tannophilus. 1985, Thesis (Doctorate in Science), Faculdade de Engenharia de Alimentos e Agricola, Universidade Estadual de Campinas, Campinas.

Gottschalk, L. M. F., Oliveira, R. A. and Bom, E. P. S. (2010). Cellulases, xylanases, β-glucosidase and ferulic acid esterase produced by Trichoderma and Aspergillus act synergistically in the hydrolysis of sugarcane bagasse. *Biochem Eng J.*, *51*, 72–78.

Grootaert, C., Delcour, J. A., Courtin, C. M., Broekaert, W. F., Verstraete, W. and Wiele, T. V. (2007). Microbial metabolism and prebiotic potency of arabinoxylan oligosaccharides in the human intestine. *Trends Food Sci Tech.*, *18*, 64–71, 2007.

Guilherme, A. A., Dantas, P. V. F., Santos, E. S., Fernandes, F. A. N. and Macedo, G. R. (2015). Evaluation of composition, characterization and enzymatic hydrolysis of pretreated sugar cane bagasse. *Braz. J. Chem. Eng.*, *32*, 23-33.

Gullón, P., Moura, P., Esteves, M., Girio, F. M., Domínguez, H. and Parajó, J. C. (2008). Assessment on the fermentability of xylo-oligosaccharides from rice husks by probiotic bacteria. *Journal of Agricultural and Food Chemistry.*, *56*, 7482–7487.

Hendriks, A. T. W. M. and Zeeman, G. (2009). Pretreatments to enhance the digestibility of lignocellulosic biomass. *Bioresour Technol.*, *100*, 10–18.

Holzapfel, W. H., Haberer, P., Snel, J, Schillinger, U. and Huis in't Veld, J. H. (1998). Overview of gut flora and probiotics. *Int J Food Microbiol.*, *41*, 85–101.

Izumi, K., Ikemizu, S. and Shizuka, F. (2004). Anti-hyperlipidemics containing acidic xylooligosaccharides. Japan Patent JP 2004182615.

Jayapal, N., Samanta, A. K., Kolte, A. P., Senani, S., Sridhar, M., Suresh, K. P. and Sampath, K. T. (2013). Value addition to sugarcane bagasse: Xylan extraction and its process optimization for xylooligosaccharides production. *Ind Crop Prods.*, *42*, 14–24.

Juhász, T., Szengyel, Z., Reczey, K., Siika-Aho, M. and Viikari, L. (2005). Characterization of cellulases and hemicellulases produced by *Trichoderma reesei* on various carbon sources. *Process Biochem.*, *40*, 3519-3525.

Kabel, M. A., Carvalheiro, F., Garrote, G., Avgerinos, E., Koukios, E., Parajó, J. C., Gírio, F. M., Schols, H. A. and Voragen, A. G. J. (2002). Hydrothermally treated xylan rich by-products yield different classes of xylo-oligosaccharides. *Carbohydr Polym.*, *47*, 47–56.

Kalogeris, E., Christakopoulos, P., Vrsanská, M., Kekos, D. and Biely, P. (2001). Catalytic properties of the endoxilanase I from *Thermoascus aurantiacus*. *J Mol Catal B: Enzym.*, *11*, 491-501.

Krishna, G., Divyashri, G., Prapulla, S. G. and Muralidhara, A. (2015). Combination supplement of fructo- and xylo-oligosaccharides significantly abrogates oxidative impairments and neurotoxicity in maternal/fetal milieu following gestational exposure to acrylamide in rat. *Neurochem Res.*, *40*, 1904–1918.

Lagaert, S., Pollet, A., Courtin, C. M. and Volckaert, G. (2014). β-Xylosidases and α-L-arabinofuranosidase: accessory enzymes for arabinoxylana degradation. *Biotechnol. Adv.*, *32*, 316-332.

Lakshmi, G. S., Rao, C. S., Rao, R. S., Hobbs, P. J. and Prakasham, R. S. (2009). Enhanced production of xylanase by a newly isolated Aspergillus terreus under solid state fermentation using palm industrial waste: A statistical optimization. *Biochem Eng J.*, *48*, 51–57.

Li, J., Kisara, K., Danielsson, S., Lindström, M. E. and Gellerstedt, G. (2007). An emproved methodology for the quantification of uronic acid units in xylans and other polysaccharides. *Carbohydr Res.*, *342*, 1442-1449.

Mäkeläinen, H., Forssten, S., Saarinen, M., Stowell, J., Rautonen, N. and Ouwehand, A. C. (2010). Xylo-oligosaccharides enhance the growth of bifidobacteria and Bifidobacterium lactis in a simulated colon model. *Benef Microbes.*, *1*, 81–91.

Manning, T. S. and Gibson, G. R. (2004). Prebiotics. *Best Pract Res Clin Gastroenterol.*, *18*, 287–298.

Maryana, R., Marifatun, R., Wheni, A., Satriyo, K. W. and Rizal, W. A. (2014). Alkaline pretreatment on sugarcane bagasse for bioethanol production. *Energy Procedia.*, *47*, 250-254.

Maslen, S. L., Goubet, F., Adam, A., Dupree, P. and Stephens, E. (2007). Structure elucidation of arabinoxylan isomers by normal phase HPLC-MALDI-TOF/TOF-MS/MS. *Carbohydr Res.*, *342*, 724-735.

Miller, J. D., Gilbert, R. A. and Odero, D. C. (2012). Sugarcane botany: A brief view. Document is SS-AGR-234, one of a series of the Agronomy Department, Florida Cooperative Extension Service, Institute of Food and Agricultural Sciences, University of Florida. Access on August of 2015: http://edis.ifas.ufl.edu/sc034.

Moniz, P., Pereira, H., Duarte, L. C. and Carvalheiro, F. (2014). Hydrothermal production and gelfil tration purification of xylo-oligosaccharides from rice straw. *Ind Crop Prod.*, *62*, 460–465.

Mosier, N., Wyman, C., Dale, B., Elander, R., Lee, Y. Y., Holtzapple, M. and Ladisch, M. (2005). Features of promising technologies for pretreatment of lignocellulosic biomass. *Bioresour Technol.*, *96*, 673–686.

Moura, P., Barata, R., Carvalheiro, F., Gírio, F., Loureiro-Dias, M. C. and Esteves M. P. (2007). *In vitro* fermentation of xylooligosaccharides from corn cobs autohydrolysis by Bifidobacterium and Lactobacillus strains. *LWT – Food Sci Technol.*, *40*, 963–72.

Moure, A., Gullón, P., Domínguez, H. and Parajó, J. C. (2006). Advances in the manufacture, purification and applications of xylo-oligosaccharides as food additives and nutraceuticals. *Process Biochem*, *41*, 1913–1923.

Mussatto, S. I. and Mancilha, I. M. (2007) Non-digestible oligosaccharides: A review. *Carbohydr Polym.*, *68*, 587–597.

Nabarlatz, D., Ebringerova, A. and Montane, D. (2007). Autohydrolysis of agricultural by-products for the production of xylo-oligosaccharides. *Carbohydr Polym.*, *69*, 20-28.

Nabarlatz, D., Farriol, X. and Montane, D. (2004). Kinetic modeling of the autohydrolysis of lignocellulosic biomass for the production of hemicellulose-derived ligosaccharides. *Ind Eng Chem Res.*, *43*, 4124-4131.

Njamela, S. N., Vena, P. F., Brienzo, M., Görgens, J. and Tyhoda, L. (2013). The effect of hemicelluloses pre-extraction on the lignin-carbohydrate complex structure of sugarcane bagasse pulp. *Tappsa Journal.*

Okazaki, M., Fujikawa, S. and Matsumoto, N. (1990). Effects of xylooligosaccharides on growth of bifidobacteria. *Bifidobacteria and Microflora.*, *9*, 77-86.

Otieno, D. O. and Ahring, B. K. (2012). A thermochemical pretreatment process to produce xylooligosaccharides (XOS), arabinooligosaccharides (AOS) and mannooligosaccharides (MOS) from lignocellulosic biomasses. *Bioresour Technol.*, *112*, 285–292.

Overend, R. P. and Chornet, E. (1989). Steam and aqueous pretreatments: are they prehydrolyses? In: Wood Processing and Utilisation. Chichester: Ellis Horwood, 395–400.

Peng, F., Ren, J. L., Xu, F., Bian, J., Peng, P. and Sun, R. C. (2009). Comparative study of hemicelluloses obtained by graded ethanol precipitation from sugarcane bagasse. *J Agric Food Chem.*, *57*, 6305-6317.

Puchart, V. and Biely, P. (2008). Simultaneous production of endo-β-1,4-xylanase and branched xylooligosaccharides by *Thermomyces lanuginosus*. *J Biotechnol.*, *137*, 34-43.

Rasmussen, H., Sørensen, H. R. and Meyer, A. S. (2014). Formation of degradation compounds from lignocellulosic biomass in the biorefinery: sugar reaction mechanisms. *Carbohydr Res.*, *385*(19), 45–57. 2014.

Rivas, B., Domínguez, J. M., Domínguez, H. and Parajó, J. C. (2002). Bioconversion of posthydrolysed autohydrolysis liquors: an alternative for xylitol production from corn cobs. *Enzyme Microb Technol.*, *31*, 431–438.

Rocha, G. J. M., Martin, C., Soares, I. B., Souto-Maior, A. M., Baudel, H. M. and Abreu, C. A. M. (2011). Dilute mixed-acid pretreatment of sugarcane bagasse for ethanol production. *Biomass and Bioenergy.*, *35*, 663-670.

Rocha, G. J. M., Gonçalves, A. R., Oliveira, B. R., Olivares, E. G. and Rossell, C. E. V. (2012). Steam explosion pretreatment reproduction and alkaline delignification reactions performed on a pilot scale with sugarcane bagasse for bioethanol production. *Industrial Crops and Products.*, *35*, 274-279.

Rocha, G. J. M., Nascimento, V. M., Gonçalves, A. R., Silva, V. F. N. and Martín, C. (2015). Influence of mixed sugarcane bagasse samples evaluated byelemental and physical–chemical composition. *Ind Crop Prods.*, *64*, 52–58.

Rose, J. (2003). The plant cell wall. Oxford: Blackwell.

Saad, M. B. W., Oliveira, L. R. M., Candido, R. G., Quintana, G., Rocha, G. J. M. and Gonçalves, A. R. (2008). Preliminary studies on fungal treatment of sugarcane straw for organosolv pulping. *Enzyme Microb Technol.*, *43*, 220-225.

Samanta, A. K., Senani, S., Kolte, A. P., Sridhar, M., Sampath, K. T., Jayapal, N. and Devi, A. (2012). Production and *in vitro* evaluation of xylooligosaccharides generated from corncobs. *Food Bioprod Process.*, *90*, 466–474, 2012.

Sant'Anna, C., De Souza, W. and Brienzo, M. (2014). The influence of the heterogeneity, physicochemical and structural properties on the recalcitrance and conversion of sugarcane bagasse. In: Sugarcane: Production, *Consumption and Agricultural Management Systems*, 1-34.

Santos, V. T., Esteves, P. J., Milagres, A. M. and Carvalho, W. (2011). Characterization of commercial cellulases and their use in the saccharification of a sugarcane bagasse sample pretreated with dilute sulfuric acid. *J Ind Microbiol Biotechnol.*, *38*, 1089-98.

Sasaki, M., Adschiri, T. and Arai, K. (2003). Fractionation of sugarcane bagasse by hydrothermal treatment. *Bioresour Technol.*, *86*, 301–304.

Saulnier, L., Vigouroux, J. and Thibault, J. F. (1995). Isolation and partial characterization of feruloylated oligosaccharides from maize bran. *Carbohydr Res.*, *272*, 241-253.

Sun, J. X., Sun, X. F., Sun, R. C. and Su, Y. Q. (2004). Fractional extraction and structural characterization of sugarcane bagasse hemicelluloses. *Carbohydr Polym.*, *56*, 195-204.

Szczerbowski, D., Pitarelo, A. P., Filho, A. Z. and Ramos, L. P. (2014). Sugarcane biomass for biorefineries: Comparative composition of carbohydrate and non-carbohydrate components of bagasse and straw. *Carbohydr Polym.*, *114*, 95-101.

Taherzadeh, M. J. and Karimi, K. (2008). Pretreatment of lignocellulosic wastes to improve ethanol and biogas production: A Review. *Int J Mol Sci.*, *9*, 1621-1651.

Takenishi, S. and Tsujisaka, Y. (1975). On the modes of action of three xylanase produced by a strain of *Aspergillus niger* van Tieghem. *Agric Biol Chem.*, *39*, 2315-2323.

Thakore, A., Oei, J., Ringrose, B., Gibson, A. and Wajer, M. (2005). The use magnesium hydroxide as a cost effective cellulose protector in the pressurized alkaline peroxide (Eop) bleaching stage. *Pulp paper Canada.*, *106*, 46-49.

Van Laere, K. M., Beldman, G. and Voragen, A. G. J. (1997). A new arabinofuranohydrolase from *Bifidobacterium adolescentis* able to remove arabinosyl residues from double-substituted xylose units in arabinoxylan. *Appl Microbiol.*, *47*, 231-235.

Vázquez, M. J., Alonso, J. L., Domínguez, H. and Parajó, J. C. (2000). Xylooligosaccharides: manufacture and applications. *Trends Food Sci Tech.*, *11*, 387-93.

Vázquez, M. J., Garrote, G., Alonso, J. L., Domínguez, H. and Parajó, J. C. (2005). Refining of autohydrolysis liquors for manufacturing xylo-oligosaccharides: Evaluation of operational strategies. *Bioresour Technol.*, *96*, 889–896.

Wang, J., Yuan, X., Sun, B., Cao, Y., Tian, Y. and Wang, C. (2009). On-line separation and structural characterization of feruloylated oligosaccharide from wheat bran using HPLC–ESI-MSn. *Food Chem.*, *115*, 1529–1541.

Weil, J., Westgate, P., Kohlmann, K. and Ladisch, M. R. (1994). Cellulose pretreatments of lignocellulosic substrates. *Enzyme Microb Technol.*, *16*, 1002-1004.

Xiao, L., Ning, J. and Xu, G. (2012). Application of Xylo-oligosaccharide in modifying human intestinal function. *Afr J Microbiol Res.*, *6*, 2116-2119.

Xu, F., Sun, J. X., Liu, F. C. and Sun, R. C. (2006). Comparative study of alkali and acid organic solvent-soluble hemicellulosic polysaccharides from sugarcane bagasse. *Carbohydr Res.*, *341*, 253-261.

Yang, B., Boussaid, A., Mansfield, S. D., Gregg, D. J. and Saddler, J. N. (2002). Fast and efficient alkaline peroxide treatment to enhance the enzymatic digestibility of steam-exploded softwood substrates. *Biotechnol Bioeng.*, *77*, 678-84.

Yang, R., Xu, S., Wang, Z. and Yang, W. (2005). Aqueous extraction of corncob xylan and production of xylooligosaccharides. *LWT-Food Sci Technol.*, *38*, 677-82.

Yu, X., Yin, J., Li, L., Luan, C., Zhang, J., Zhao, C. and Li, S. (2015). Prebiotic potential of xylooligosaccharides derived from corn cobs and their *in vitro* antioxidant activity when combined with lactobacillus. *J Microbiol Biotechnol.*, *25*, 1084–1092.

Yuan, X., Wang, J. and Yao, H. (2004). Antioxidant activity of feruloylated oligosaccharides from wheat bran. *Food Chem.*, *90*, 759-764.

Zhao, X., Peng, F., Cheng, K. and Liu, D. (2009). Enhancement of the enzymatic digestibility of sugarcane bagasse by alkali-peracetic acid pretreatment. *Enzyme Microb Technol.*, *44*, 17-23.

INDEX

#

5-hydroxymethylfurfural, 171

A

access, 4, 10, 61, 151
accessibility, xii, 156, 171
accounting, 96, 149, 151
acetic acid, 76, 172, 174
acetoclastic methanogens, 76
acetogenesis, 74, 76, 80, 100, 101
acid, xi, 27, 29, 30, 31, 35, 38, 41, 48, 55, 56, 57, 60, 61, 67, 76, 85, 138, 142, 145, 148, 149, 156, 157, 164, 170, 171, 172, 173, 174, 178, 180, 183, 184, 186, 187, 188
acid, 170, 171, 177
acid esterase, 177
acid hydrolysis, 156, 171, 173
acidic, xii, 41, 44, 47, 59, 156, 167, 172, 183
acidogenesis, 74, 75, 80, 81, 92, 100, 117
activated carbon, 31, 157
adaptation, 22, 133
additives, xii, 24, 43, 156, 168
adhesion, 169
adsorption, 31, 65, 158, 174
adults, 126
adverse effects, 24
Africa, 3, 8, 9, 10, 12
age, 83, 124, 126, 161
aggression, viii, 23
aging process, viii, 23
agricultural economics, 18
agriculture, 4, 5, 8, 13, 15, 18, 20, 21, 86, 157
alcohols, x, 72, 76, 87, 100, 171
alkaline, xii, 38, 60, 61, 83, 85, 86, 156, 163, 164, 165, 170, 174, 175, 176, 180, 184, 186, 187, 188
alkalinity, 77, 79, 103
alkaloids, 41
allergy, 178
α-glucuronidase, 177
alternative energy, 138
aluminium, 140
amino, 39, 54, 74, 100
amino acid, 39, 54, 74, 100
ammonia, 79, 114, 116, 119
anaerobic bacteria, 74, 77, 79, 80, 84, 154
anaerobic digestion, vii, x, 27, 71, 72, 74, 75, 76, 77, 78, 79, 80, 81, 83, 84, 87, 88, 89, 90, 91, 92, 93, 96, 100, 101, 103, 105, 106, 112, 114, 115, 116, 117, 118, 119
anaerobic digestion, 74, 89, 100, 114, 116
Anaerobic Sequencing Batch Reactors, 81, 82
anaerobic sludge, 91
animal feeding operations, 117

anthocyanin, 34, 47
antibiotic, 158
anti-cancer, 178
anticancer activity, 47
antioxidant, vii, viii, ix, 23, 24, 26, 28, 29, 30, 31, 32, 34, 35, 36, 37, 38, 39, 40, 41, 42, 43, 44, 45, 46, 47, 48, 49, 52, 56, 57, 58, 59, 61, 62, 63, 64, 65, 66, 67, 68, 169, 178, 188
apples, 128
aquaculture, 27
aqueous solutions, 36, 56
arabinofuranosidase, 177, 184
arabinogalactan, 54
archaea methanogenic, 7, 776, 79
ascorbic acid, 42, 59
Asia, 3, 8, 9, 159
Asian countries, 4, 158
Aspen Plus, 96, 106, 114, 117, 119
Aspergillus terreus, 184
assessment, 20, 67, 68, 89, 114, 116, 139
atmosphere, x, 95, 99, 102, 145, 146
attribution, 34
Austria, 107, 119, 127, 129, 131
autohydrolysis, 156, 157, 164, 169, 172, 181, 182, 185, 186, 187

B

Bacillus subtilis, 32, 182
bacteria, 33, 60, 74, 75, 78, 79, 80, 85, 100, 101, 161, 167, 168, 169, 170, 179, 182, 183
bargaining, viii, 2, 6
barriers, 123, 133, 136, 176
base, 5, 12, 86, 88, 89, 159
batteries, 97
beef, 38, 44
beer, 38, 47
Belgium, 11
beneficial effect, 168
benefits, viii, ix, 23, 25, 42, 43, 47, 52, 53, 63, 102, 103, 105, 114, 158, 168, 181
beverages, ix, 25, 49, 51, 53
bicarbonate, 79
bifidobacteria, 158, 178, 182, 184, 185
bioavailability, 77, 92
bioconversion, 176
biodegradability, 41, 89
biodegradable materials, 74
biodegradable wastes, 103
biodegradation, xii, 73, 156
biodiesel, 43, 67, 69, 73, 83, 86
bioenergy, 70, 74, 85, 86, 87, 89, 92, 93, 114, 157
biofuels, ix, x, 13, 51, 53, 66, 72, 74, 83, 86, 87, 90, 95, 112, 156, 157
biogas, x, 72, 73, 74, 76, 77, 78, 79, 80, 81, 83, 84, 86, 89, 92, 93, 95, 96, 100, 103, 104, 105, 106, 107, 110, 111, 112, 115, 116, 117, 119, 187
biological activity, 29, 50
biological processes, 87, 89
biological systems, 24
biologically active compounds, 53
biomass, viii, 23, 67, 69, 85, 86, 87, 101, 102, 115, 138, 154, 157, 159, 160, 161, 163, 164, 171, 172, 173, 174, 180, 183, 184, 185, 187
biomolecules, 30
biopolymers, x, 39, 72
bio-products, 86
biorefineries, 27, 86, 87, 89, 92, 185
biorefinery chain, 87, 88
bleaching, 175, 187
blood, 30, 158
boilers, ix, 51, 52
branching, 170, 177
Brazil, 47, 67, 69, 71, 78, 155, 157, 159
breakdown, 100, 172, 173, 174, 175
breeding, 27, 160
by-products, vii, viii, xi, 24, 25, 34, 35, 37, 40, 44, 45, 46, 49, 65, 66, 67, 75, 137, 170, 171, 173, 174, 176, 178, 183, 185

C

C/N ratio, 79, 82, 83, 86, 105
cabbage, 85
CAF, 117

caffeine, 33, 53, 54, 55, 57, 58, 61, 63, 66, 67
calcium, 158, 161
calibration, 61, 142
calorie, 158, 168
calorimetry, 144, 145
campaigns, 12, 13, 123, 124, 125, 133
cancer, 24, 46, 53, 68
candidates, 35
carbohydrate(s), 73, 74, 85, 116, 164, 174, 182, 185, 187
carbon, 29, 41, 47, 48, 59, 72, 74, 76, 77, 79, 81, 101, 102, 103, 105, 116, 147, 148, 151, 152, 157, 159, 169, 183
carbon atoms, 159
carbon dioxide, 29, 72, 101, 102, 103, 147, 148, 151, 152
carbon monoxide, 147, 152
carboxylic acid, 164
carboxymethyl cellulose, 157
carcinoma, 29, 30
cardiovascular disease, 24, 46, 158, 168
cardiovascular system, 53
carnosine, 36
carotene, 32, 35, 36, 39
carotenoids, 35, 36, 37, 39, 42
case study, 45
catalysis, 171
catalyst, 164, 172
cell line, 67
cellulose, xi, 54, 55, 86, 153, 156, 157, 159, 160, 161, 162, 164, 171, 174, 176, 187
Central Asia, 12
cephalothorax, 39
cheese, 128
chemical characteristics, 84
chemical properties, 36
chemicals, 26, 83, 87, 172, 179
chemiluminescence, 153
chemistry composition, 74
chemometrics, 68
chicken, 81, 168
China, 8, 10, 103, 104, 115, 116, 119, 159, 170
chitosan, 39
chlorophyll, 36
cholesterol, 36, 55
chromatographic technique, 20
chromatography, vii, xi, 33, 36, 137
circulation, 149
clean energy, x, 72
cleaning, 97, 114
cleavage, 172, 177
cleavages, 167
climate change, 122
climates, 111, 117
CO_2, 9, 55, 62, 66, 68, 75, 76, 79, 80, 142, 159
coal, 48
coatings, 39
co-digestion processes, 76
coffee, ix, 25, 38, 41, 46, 51, 52, 53, 54, 55, 56, 57, 58, 59, 60, 61, 62, 63, 64, 65, 66, 67, 68, 69, 70, 74
cogeneration, 84
colon, 29, 30, 158, 168, 169, 179, 182, 184
colon cancer, 158, 168
combined effect, 56
combustion, 52, 66, 84, 143, 161
commercial, ix, 27, 39, 51, 52, 59, 65, 74, 86, 87, 89, 97, 110, 114, 170, 186
commodity, ix, 51
common agricultural policy, 17, 18
composition, x, xi, 30, 31, 32, 34, 37, 49, 50, 53, 54, 56, 59, 60, 61, 62, 63, 64, 65, 68, 73, 74, 82, 83, 85, 87, 95, 105, 135, 138, 156, 157, 159, 160, 161, 162, 165, 166, 167, 171, 183, 186, 187
compost, 66, 99
composting, 53, 66, 99, 102, 103, 105, 112, 116
compounds, vii, viii, ix, 23, 24, 26, 27, 28, 31, 32, 35, 36, 37, 39, 41, 42, 43, 46, 47, 49, 52, 53, 54, 55, 56, 57, 58, 60, 61, 63, 64, 65, 67, 77, 80, 100, 158, 161, 168, 171, 174, 175, 179, 185
condensation, 173
conductivity, 142
configuration, 81, 82
congruence, 13

conservation, 17, 154
constipation, 170
constituents, 41, 44, 66, 68, 160
consumer protection, 20
consumers, viii, 1, 3, 6, 17, 19, 122, 124, 127, 130, 133, 136, 158, 179
consumption, vii, xi, 2, 3, 4, 8, 9, 10, 11, 15, 20, 21, 24, 43, 53, 64, 68, 118, 121, 122, 124, 129, 133, 134, 146, 158, 168, 169, 170
consumption patterns, 133
contamination, 99
convergence, 12, 13
conversion rate, 101
cooking, 3, 9, 39, 124, 125, 127, 128, 131, 132, 133
cooling, 111, 112
cooperation, 11, 12, 13
copper, 30
correlation, 29, 31, 37, 53, 59
corrosion, 41, 43, 44, 46, 48, 171, 172
cosmetics, 30, 60, 87
cost, x, xii, 3, 4, 6, 32, 72, 84, 89, 99, 104, 156, 158, 174, 176, 187
Croatia, 89, 91
crops, 7, 8, 16, 86
crystalline, 160, 176
crystallinity, 174
cultivation, ix, 51
culture, 11, 85, 127, 169, 176
culture media, 169
culture medium, 169, 176
cycles, 25, 161
cyclooxygenase, 66
cytometry, 29
cytotoxicity, 30

D

danger, 147, 152, 153
DDGS, 115
deaths, xi, 137
decay, 128, 168
decision makers, xi, 96
decomposition, 24, 99, 143, 144, 145, 153, 175
decomposition temperature, 143, 144
deficiency, xi, 79, 137, 138, 148, 152
degradation, viii, 23, 25, 33, 40, 43, 72, 81, 92, 103, 104, 152, 171, 172, 173, 174, 175, 179, 184, 185
degradation rate, 40
dehydration, 143
Delta, 117
Denmark, 7
Department of Agriculture, 3, 15
dependent variable, 133
depolymerization, 54, 172, 175, 179
deposits, xi, 138, 152
depression, 53, 66, 67
deprivation, 17, 139
derivatives, xii, 24, 27, 38, 44, 45, 156
desorption, 31
destruction, 99
detection, 61, 68
developed countries, xi, 7, 10, 11, 12, 121, 132
developing brain, 169
developing countries, vii, viii, x, xi, 1, 3, 7, 9, 10, 11, 12, 13, 72, 104, 121, 122
developing economies, 122
development policy, 12
diabetes, 53, 168
diarrhea, 170
diet, 10, 24, 126, 158
dietary fiber, ix, 26, 32, 35, 44, 52, 55, 62, 63, 64, 65, 67, 69
diffuse reflectance, 68
digestibility, 76, 161, 168, 183, 188
digestion, x, 73, 74, 76, 78, 79, 80, 82, 83, 85, 89, 90, 91, 96, 100, 102, 103, 105, 106, 112, 114, 115, 116, 117, 118
digestive enzymes, 168
dioxin, 99
diseases, viii, ix, 23, 24, 52, 53
dissociation, 172
distilled water, 147, 150, 151, 152
distortions, viii, 1, 4, 6, 7, 12, 15

distribution, viii, 1, 2, 5, 6, 7, 9, 12, 15, 85, 87, 122, 123, 138, 159, 160
diversity, 30, 91, 157
DNA, viii, 23, 36
dosage, 169, 178
drainage, 119
dry matter, 123
drying, 27, 66, 139
DTA curve, 143
dumping, x, 71

E

E. coli, 170
East Asia, 3, 9, 12
Eastern Europe, 12
economic development, 5, 14
economic growth, 14, 15, 96, 113
Economic Research Service, 16
economic systems, 5
ecosystem, 25, 180
education, 126, 131
effluent, 45, 73, 77, 82, 100
electricity, x, 73, 84, 86, 95, 99, 106, 110, 111, 112, 157, 161
electrochemical impedance, 42
electron, 30
elucidation, 184
emission, 10, 99, 100, 142
employment, 32, 126
endothermic, 143
endo-β-1, 4-D-xylanase, 177
energy, x, xii, 10, 25, 47, 72, 73, 74, 77, 78, 79, 83, 84, 85, 86, 88, 89, 91, 92, 95, 96, 102, 103, 104, 105, 106, 110, 111, 112, 114, 115, 116, 117, 138, 153, 156, 171, 179
energy consumption, 104, 110
energy recovery, x, 25, 95, 106
energy supply, 10, 114
environment, x, xi, 5, 9, 14, 16, 17, 43, 72, 74, 95, 100, 138
environmental, viii, ix, x, xi, 2, 4, 15, 18, 20, 21, 22, 25, 51, 52, 71, 72, 77, 81, 91, 92, 96, 99, 103, 105, 106, 112, 113, 114, 121, 122, 132, 133, 138, 156, 161, 176, 179
environmental aspects, 105
environmental conditions, 77, 81
environmental impact, ix, 2, 25, 51, 122, 176
environmental issues, 99, 114
environmental protection, 179
environmental sustainability, viii, 2, 4, 15, 22, 138
enzymatic hydrolysis, 65, 157, 170, 171, 175, 176, 177, 178, 179, 180, 181, 183
enzymes, x, 27, 72, 74, 77, 80, 83, 86, 10, 157, 160, 1670, 176, 177, 184
EOG, 139, 144, 145
epithelium, 169
erosion, 17
ESI, 187
essential fatty acids, 35
ester, 148, 149, 153, 164, 172, 174
ethanol, xi, 28, 29, 30, 31, 32, 33, 36, 37, 38, 39, 46, 57, 58, 59, 60, 62, 69, 72, 73, 76, 83, 86, 87, 89, 155, 157, 161, 165, 180, 185, 186, 187
ethyl acetate, 29, 30
ethylene, 139
ethylene oxide, 139
Europe, 3, 4, 6, 32, 78
European Commission, 19, 104, 115
European Parliament, 115
European Union, 12, 13, 17, 19, 115, 134
evaporation, 30
evolution, viii, 1, 4, 14, 15, 17, 19, 21
exo-β-1,4-xylanase, 177
experimental design, 46
exploitation, 5, 8, 20, 21, 25, 39, 43
exposure, 184
extraction, vii, xii, 28, 29, 30, 32, 33, 35, 36, 38, 41, 43, 44, 46, 53, 54, 57, 58, 59, 60, 62, 63, 64, 68, 70, 87, 89, 139, 156, 157, 159, 160, 161, 163, 164, 165, 166, 167, 170, 174, 175, 176, 177, 178, 179, 180, 183, 185, 187, 188

extracts, 27, 28, 29, 30, 32, 36, 37, 38, 39, 41, 43, 45, 46, 47, 48, 57, 58, 60, 61, 62, 65, 67, 178
extrusion, 41

F

family members, 132
farmers, viii, 1, 6, 11
fat, 34
fatty acids, 35, 60, 74, 75, 78, 79, 100, 158, 161, 170
feedstock, 47, 84, 87, 89, 118, 119, 156, 159, 161, 162, 172, 179
fermentation, xi, xii, 32, 33, 81, 82, 83, 85, 87, 88, 89, 90, 91, 92, 116, 138, 139, 141, 144, 145, 146, 147, 148, 149, 151, 153, 156, 157, 169, 181, 182, 184, 185
fermentative bacteria, 74, 75
fertilization, 157
fertilizers, ix, 9, 51, 52, 87
fetus, 168
fiber, 35, 55, 56, 62, 63, 157, 159
fiber content, 35, 63
fibers, vii, ix, 26, 52, 55, 63, 160, 167
filtration, 27, 29
financial support, 43, 63
Finland, 107
fires, 154
fish, vii, xi, 4, 36, 74, 84, 137, 138, 139, 143, 144, 145, 146, 147, 149, 150, 151, 152
fisheries, 2
flavonoids, 29, 36, 37, 41, 61
flora, 183
flour, 36, 49, 55, 69
flowers, 37
fluid, 29, 62, 64
fluid extract, 29, 62, 64
food additives, viii, 23, 24, 185
food chain, xi, 121
food industry, vii, ix, xi, 43, 45, 52, 137, 138
food production, vii, xi, 9, 13, 25, 121
food products, viii, 2, 6, 7, 10, 23, 24, 25, 26, 43, 60, 123, 131
food safety, 3, 10
food security, 10, 11, 25
formation, 8, 24, 34, 41, 56, 76, 79, 102, 158, 167, 170, 171, 172, 175, 176
formula, 108
France, 6, 140
free radicals, viii, 23, 24, 34, 175
freshwater, 31, 135
fruits, 25, 37, 55, 74, 84, 128
FTIR, 59
fuel cell, 87, 110
fuel consumption, 112
functional food, viii, ix, 23, 25, 26, 30, 32, 33, 35, 36, 42, 52, 63, 158, 179
fungi, 161
furan, 99
furfural, 171, 173, 179

G

garbage, 92, 154
gasification, 72
gastrointestinal tract, 168
gel, 164
genus, 159, 168, 169, 170
Germany, 115
gestation, 168
ginseng, 32, 47
glucose, 54, 59, 74, 86, 160, 162, 163, 166
glutamate, 67
glycerol, 32, 36
grass, 117, 159
Great Britain, 6
Greece, v, vii, xi, 121, 122, 123, 125, 126, 127, 131, 132, 133, 134
greenhouse gas (GHG), 9, 10, 15, 89, 99, 102, 104, 105
greenhouse gas emissions, 15, 102, 104
groundwater, 99
growth, ix, 13, 33, 60, 67, 71, 96, 161, 168, 169, 170, 176, 178, 179, 181, 184, 185
guilt, 129

H

harvesting, 37, 157
hazardous waste, 97
hazards, 19, 138, 153
health, viii, ix, 23, 24, 36, 37, 42, 43, 47, 49, 50, 52, 53, 59, 63, 64, 66, 99, 158, 168, 179, 181
health effects, 24, 43, 49, 53, 59, 99
health promotion, 36
heat loss, 110
heat transfer, 108
heating rate, 140
heavy metals, 80, 99, 175
hemicellulose, vii, xii, 54, 55, 156, 157, 159, 160, 161, 163, 165, 170, 171, 172, 174, 176, 177, 178, 179, 180, 185
hemicellulose extraction, xii, 156, 161, 170, 177, 179, 180
hemicellulose hydrolysis, 171, 178
heterogeneity, 157, 173, 177, 186
hexane, 55, 161
homogeneity, 87
human attitudes, 133
human body, 24
human health, 100, 158
humidity, 147
hybrid, 80, 124
hydrogen, x, 55, 72, 76, 78, 82, 83, 85, 86, 87, 88, 89, 90, 91, 100, 101, 147, 160, 175
hydrogen bonds, 160
hydrogen peroxide, 55, 175
hydrogen production, 85, 91
hydrolysis, 61, 65, 74, 75, 80, 81, 86, 100, 156, 157, 170, 171, 173, 175, 176, 177, 178, 179, 180, 181, 183
hydrosphere, x, 95
hydrothermal pretreatment, 172, 174
hydrothermal process, 179
hydroxide, 187
hydroxyl, 28, 36, 61, 164, 175
hydroxyl groups, 164
hypotensive, 69

I

immersion, 42
immune system, 168
immunomodulatory, 178
immunostimulatory, 55
impairments, 184
improvements, 168
impurities, 108, 164
in vitro, 37, 68, 69, 168, 169, 170, 186, 188
in vivo, 168, 169, 170
incidence, 24
India, 8, 103, 104, 105, 107, 117, 159
individuals, 11
industrial processing, 35
industrial transformation, 9
industrial wastes, 86, 87, 102
industrialized countries, viii, 1, 3
inflammatory responses, 66
infrastructure, 123
ingredients, viii, ix, 23, 25, 26, 35, 42, 52, 63, 86, 138, 158
inhibition, 34, 36, 41, 44, 47, 48, 59, 60, 79, 80, 103, 116
inhibitor, 46
inoculation, 66
inoculum, 83, 84
institutions, 6, 97, 114
intensive farming, 8
International Energy Agency (IEA), 86
internode, 159, 160, 161
intervention, 136
intestinal flora, 168
intestinal tract, 167
intestine, 74, 168, 170, 183
investments, 12, 13
ionization, 77
Ireland, 91, 117
iron, 77
irrigation, 8
isolation, 159, 164, 177
isomerization, 34
isomers, 24, 35, 184
Italy, 1, 8, 18, 19, 20, 21

J

Japan, 12, 137, 138, 140, 142, 153, 158, 169, 183

K

ketones, 171
kinetic studies, 117
kinetics, 46, 66, 89
KOH, 174
Korea, 12, 105

L

lactic acid, 85, 100, 169
Lactobacillus, 75, 168, 169, 170, 185, 188
landfills, 25, 72, 74, 83, 88, 100
Latin America, 9, 12
LDL, 27, 30, 36
LIFE, 134
light scattering, 164
lignans, 161
lignin, xii, 34, 54, 55, 59, 156, 157, 159, 160, 161, 162, 163, 164, 166, 170, 171, 173, 174, 175, 176, 179, 180, 185
lignin-carbohydrate-complex, 164
lignocellulosic materials, 62, 176, 181
Likert scale, 132
linoleic acid, 35
lipid oxidation, 35, 39, 41, 43
lipids, 59, 73, 100
liquid chromatography, 33, 41, 78
Listeria monocytogenes, 60
livestock, 69, 168
local authorities, 133
low density polyethylene, 48
low temperatures, 149
lycopene, 39

M

macromolecules, 61, 63, 157, 159, 173
macronutrients, 9
macrophages, 66
magnesium, 161, 165, 166, 175, 187
Maillard reaction, 56, 57, 61
majority, 42, 43, 57, 63, 126, 129, 147
Malaysia, 84, 125
MALDI, 184
malnutrition, viii, 2, 7, 10, 11, 12, 15, 122
management, vii, xi, 27, 96, 102, 103, 106, 112, 113, 116, 117, 131
manganese, 161
manufacturing, 92, 122, 123, 187
manure, 66, 85, 102, 104, 116
marital status, 124
market share, 9
mass, x, 36, 37, 41, 48, 71, 74, 89, 106, 114, 123, 143, 144, 159, 160, 162, 164, 165, 166
mass loss, 143, 144
mass spectrometry, 36, 37
materials, 25, 33, 38, 43, 62, 74, 75, 83, 85, 87, 100, 101, 103, 104, 123, 138, 139, 153, 157, 176, 181
matrix, 40, 43, 54, 161
measurement, 142, 144
measurements, 124, 140, 142, 145
meat, ix, 4, 10, 71, 74, 81, 84, 128, 158
media, 16, 41
medicine, 22, 46
melon, 35
membranes, 30
metabolism, 183
metabolites, viii, 23, 24, 69
metals, 175
methane, x, 25, 72, 74, 76, 77, 78, 79, 81, 83, 84, 87, 88, 89, 91, 100, 101, 102, 103, 105, 111, 115, 116, 152
methane production, 84
methanogenesis, 74, 80, 81, 100, 102
methanol, 29, 30, 56, 57
methodology, 46, 47, 58, 59, 70, 116, 135, 184
methylene blue, 65
microbiota, 158, 168, 169, 181, 182
micronutrients, 10

microorganisms, ix, 33, 67, 71, 75, 76, 77, 85, 103, 157, 158, 168, 169, 176
microwave radiation, 59
migration, 41
Miscanthus, 169, 181
missions, 9, 25
mixing, 77, 103, 107, 108
model system, 36, 68
modelling, 117, 119
models, viii, 1, 5, 7, 14, 15, 17, 21, 61, 169, 181
moisture, x, 3, 71, 105, 116, 146, 147, 149, 151
moisture content, 105, 116, 146, 147, 149
mold, 3
molecular mass, 29, 164
molecular structure, 181
molecular weight, 27, 28, 31, 45
molecules, 24, 31, 41, 156, 164, 167
monomers, 171
monosaccharide, 59, 170, 171
Moon, 47
MSW, x, xi, 91, 95, 96, 97, 98, 99, 100, 104, 105, 106, 112, 113, 114, 115, 117
municipal solid waste, xi, 72, 78, 90, 96, 116, 117, 119, 135
mutagen, 60

N

natural gas, 110
natural resources, 9, 135
near infrared spectroscopy, 61
negative effects, 158
nervous system, 53
Netherlands, 7, 22, 104
neurons, 67
neurotoxicity, 67, 184
neutral, 55
NIR, 61, 67, 68
NIR spectra, 61
nitrogen, 9, 24, 53, 54, 59, 77, 79, 103, 105, 145, 146, 153
nitrous oxide, 102
Nobel Prize, 10
nodes, 159
North Africa, 12
North America, 3, 8, 12
Norway, 6, 16
nutraceutical, 30, 56
nutrient, 2, 24, 25, 76, 79, 80, 82, 83, 84, 99, 102, 103, 104, 158, 168
nutrition, 10, 87

O

obesity, 10, 11, 122
OECD, 9, 13, 16
oil, 4, 26, 30, 35, 59, 62, 73, 86, 139
oil samples, 35
oligomers, xii, 156, 167, 171
oligopolies, 5, 6, 7
oligosaccharide, 167, 187
olive oil, 20, 26, 30, 47
operations, 73, 85, 116
opportunities, 45, 179
optimization, 58, 89, 178, 183, 184
organic compounds, 80, 159
organic matter, 82, 88, 99, 100, 102, 104
organism, 158
organs, 74
osmosis, 27
overweight, 10, 11
oxidation, xi, 24, 27, 30, 34, 36, 39, 41, 72, 100, 103, 138, 142, 145, 148, 149, 153
oxidation products, 24
oxidative agents, 174
oxidative damage, 28, 43
oxidative stress, ix, 29, 30, 52, 169
oxygen, vii, ix, xi, 24, 25, 51, 52, 74, 75, 100, 137, 138, 139, 146, 147, 148, 150, 151, 152, 171, 175

P

parenchyma, 159
pathogens, viii, 23, 99, 158
peptide, 36
permeation, 164

peroxide, 24, 69, 163, 164, 175, 180, 187, 188
pests, 2, 3
petroleum, 86
pH, 29, 31, 33, 75, 77, 78, 79, 80, 81, 82, 83, 86, 103, 108, 167, 170, 172, 173, 175
pharmaceutical, 60, 87, 179
phenol, 28
phenolic compounds, 25, 26, 28, 30, 31, 34, 35, 36, 38, 41, 43, 44, 56, 58, 61, 67, 68, 70, 171
phosphate, 74
phosphorous, 9, 161
physical properties, 181
physicochemical properties, 159, 181
phytosterols, 35
pigs, 69
pith, 159
placebo, 170
plants, viii, ix, 23, 24, 32, 71, 104, 112, 119, 157, 181
plastics, x, 72, 83
PLS, 61
polar, 37
policy, 18, 104, 123, 136, 138, 153
pollutant, ix, 9, 51, 52, 99
pollution, 72
polymers, 24, 5, 1605, 182
polymeric materials, 43
polymerization, 29, 158, 164, 170, 172, 178
polyphenols, vii, viii, 23, 24, 26, 27, 31, 32, 33, 35, 36, 37, 38, 41, 43, 44, 45, 46, 47, 48, 49, 59, 61
polysaccharides, xii, 42, 54, 55, 59, 69, 100, 156, 157, 158, 159, 160, 163, 170, 171, 175, 177, 184, 187
polyunsaturated fat, 60
polyunsaturated fatty acids, 60
population, ix, 5, 7, 10, 13, 25, 71, 79, 96, 97, 98, 103, 106, 125, 133, 156, 168
population growth, 98, 156
porosity, 55
positive correlation, 32, 43, 58
potassium, 9, 161
potato, 86
poultry, 66, 74
prebiotic, xii, 55, 62, 156, 157, 159, 168, 169, 178, 179, 181, 183
precedents, 133
precipitation, 42, 165, 185
pregnancy, 168
prevention, vii, xi, 13, 16, 18, 35, 43, 46, 121, 122, 123, 124, 125, 126, 127, 129, 133, 134, 135, 136, 168
price taker, 6
primary products, 3
private investment, 13
private sector, 133
probe, 29
probiotic, 168, 169, 182, 183
process control, 82
production costs, 176
profit, 5
proliferation, 146
protection, ix, 17, 34, 41, 43, 52, 60, 158, 161
protein hydrolysates, 36
protein oxidation, 41
proteins, 36, 39, 46, 79, 100
public concern, 114
pulp, 33, 55, 61, 74, 86, 175, 185
purification, 30, 43, 85, 171, 174, 184, 185
purity, 30, 33
PVP, 47
pyrolysis, 56, 72

Q

quality control, 82
quantification, 4, 26, 123, 133, 135, 184
questionnaire, xi, 122, 125, 127, 131

R

radiation, viii, 23, 59
radicals, 24, 28, 46, 175
raw materials, x, 13, 45, 71
reaction mechanism, 185
reaction time, 172, 173, 178

reactions, 76, 173, 176, 182, 186
reactive oxygen, 30
reactivity, 163, 170
Reactors Configurations, 80, 82
recovery, ix, 25, 26, 27, 28, 29, 31, 32, 33, 39, 41, 42, 43, 44, 45, 46, 47, 48, 51, 56, 58, 61, 70, 81, 92, 133, 170, 171, 181
recycle, 88
recycling, vii, xi, 76, 137, 138, 153
red blood cells, 30
red wine, 38, 56
reduction, 11, 16, 17, 88, 102
refinery, 88
renewable energy, 76, 84, 103, 105, 112, 119
renewable fuel, 102
residue, vii, xi, 41, 46, 52, 62, 72, 74, 88, 137, 138, 143
residues, viii, ix, x, 24, 33, 47, 48, 49, 51, 52, 53, 56, 61, 62, 63, 65, 69, 70, 72, 97, 102, 129, 158, 163, 167, 176, 187
resistance, 160, 161, 167, 168
resource management, 119
resources, ix, xi, 5, 7, 8, 13, 14, 15, 20, 21, 71, 74, 88, 115, 135, 156
restaurants, ix, 71, 87
restrictions, 25
retail, 3, 4, 6, 10, 11, 17, 97, 122, 123
reuse, 88
reverse osmosis, 27
rice field, 8
rice husk, 183
rings, xi, 121
risk, ix, x, 24, 52, 67, 68, 95, 99, 139, 152, 158, 168
risk assessment, 139
risk factors, 24
Romania, 124
room temperature, 38, 140, 143, 152, 161, 174
roots, 4, 37
routes, xi, 87, 102, 136, 155, 159
routines, 124, 127, 128, 136
royal society, 135
rural areas, x, 72
rural development, 18
Russia, 8

S

safety, 138, 147, 152
salmon, 40, 48, 49
Salmonella, 168
saturation, 33
savings, 122
scanning electron microscopy, 42
scarcity, x, 10, 95
scientific publications, 21
sclerenchyma, 159
seafood, 4
Second World, 4
secondary education, 126
seed, 5, 34, 35, 36, 40, 49, 84, 85
selectivity, 33, 36, 178
sensitivity, 142
sequencing, 83, 90
sewage, 82, 85, 90, 92, 116, 118
shelf life, viii, 23, 24, 33, 38, 41
short supply, 80
shrimp, 39, 40, 46
sialic acid, 32
side effects, 170
silica, 47, 161
silver, 38, 48, 61
simulation, 91, 119
single stage reactor, 80
skin, 33, 36, 38, 48, 49, 61, 158
sludge, 66, 73, 81, 82, 84, 85, 87, 90, 115, 116
social costs, viii, 2
social problems, 2
sodium, 161
soil erosion, 17
solid matrix, 57, 60, 63
solid state, 80, 115, 184
solid waste, 32, 46, 47, 53, 65, 72, 74, 77, 91, 96, 97, 99, 104, 113, 115, 116
solvents, 29, 32, 36, 44, 55, 63, 161
South Africa, 135
sovereignty, 7

spectroscopy, 42, 67, 68
squamous cell carcinoma, 67
stability, viii, x, 10, 23, 28, 33, 38, 40, 44, 46, 62, 87, 95, 167
standard deviation, 129
starch, 35, 44, 86
steel, 41, 43, 44, 46, 47, 48
sterilisation, 153
stimulation, 182
stomach, 157
storage, vii, xi, 28, 39, 40, 48, 52, 83, 85, 110, 122, 124, 128, 130, 137, 138, 139, 142, 147, 148, 150, 151, 152, 153, 154, 159, 161
strontium, 161
structural characteristics, 62
structure, xii, 54, 69, 80, 156, 157, 158, 159, 160, 164, 167, 174, 176, 181, 185
substitution, 48, 170
substrate, 24, 51, 67, 74, 77, 79, 82, 87, 101, 103, 105, 167, 176, 178, 187, 188
sucrose, 157, 159
sugar beet, 86
sugarcane, vii, xi, 35, 90, 155, 156, 157, 159, 160, 162, 163, 165, 180, 181, 182, 183, 184, 185, 186, 187, 188
Sugarcane bagasse, 157, 162, 163, 165
sulfate, 165, 166
sulfuric acid, 180, 181, 186
supplementation, 168
supply chain, viii, 1, 2, 3, 4, 5, 6, 7, 9, 11, 12, 15, 18, 122, 135
surface area, 171, 174
surfactant, 27, 81
surplus, 110, 112
sustainability, 7, 8, 9, 13
sustainable development, 12, 21, 67, 116, 104
sustainable energy, 117
Sweden, 78, 141
sweeteners, 37, 158, 168
symbiosis, 87
synergistic effect, 42

T

tannins, 41, 59
temperature, xi, 33, 58, 77, 78, 82, 90, 91, 93, 102, 103, 108, 110, 111, 138, 140, 142, 144, 145, 147, 150, 151, 152, 154, 172, 173, 178
territorial, 20, 21, 87
Thailand, 159
thermal analysis, vii, xi, 137, 138, 139
thermal decomposition, 143
thermal energy, 138
Third World, 17
tocopherols, 35
tooth, 168
total product, 4
tourism, 20, 21
toxic gases, 72
toxicity, 32, 77, 79, 103
trace elements, 80, 103, 118
transference, 171
transformation, 87
transportation, 99, 111, 112, 122
trial, 115
two-stage reactor, 80

U

ultrasound, 32, 46, 62
United Nations, 12, 72
Upflow Anaerobic Sludge Blanket, 81
urban areas, xi, 96, 97, 98, 99, 106, 111, 113, 117, 119, 126
urban population, 96, 126
urbanisation, x, 95, 96, 97, 106, 113
USDA, 3, 4, 17
UV radiation, 34

V

vacuum, 38
validation, 89
valorization, vii, x, 27, 68, 72, 83, 92
variables, 64, 172

variations, 55, 78, 81, 84
varieties, 32, 36, 160, 180, 181
vascular bundle, 159
vegetables, 4, 8, 25, 26, 65, 74, 81, 84, 128
ventilation, 147
Vietnam, v, vii, x, xi, 95, 96, 97, 98, 99, 100, 103, 104, 106, 108, 110, 111, 112, 113, 114, 116, 117, 118, 119
vinasse, 81, 90
viscosity, 55
vision, 12
vitamin C, 36
vitamin E, 30
vitamins, 74
volatile fatty acids, 75, 77, 78, 79, 82, 100
volatile organic compounds, 25

W

waste disposal, 25, 99, 100, 112, 114
waste heat, 84
waste management, x, 72, 73, 99, 104, 105, 106, 115, 116, 117, 133, 138
waste treatment, 89, 90, 103, 116
wastewater, ix, 26, 28, 30, 44, 45, 46, 51, 66, 74, 81, 82, 87, 92, 117, 118, 119
water, 7, 8, 15, 25, 26, 28, 29, 31, 32, 36, 38, 46, 52, 53, 54, 55, 56, 57, 58, 60, 61, 62, 63, 70, 73, 79, 85, 116, 119, 122, 139, 140, 146, 148, 153, 161, 163, 164, 165, 166, 167, 171, 172, 174, 180, 182
water evaporation, 140
weight loss, 143
welfare, 14
wood, 67, 97, 153, 182
workers, xi, 56, 137, 138
working groups, 20
World Bank, 111, 116, 118
worldwide, vii, ix, x, xi, 71, 96, 107, 121, 122, 133, 157, 168, 179

X

Xylooligosaccharides, vi, 155, 157, 167, 169, 180, 187

Y

yeast, 60